DATE DUE

DEMCO 38-296

Japan Business and Economic Series

This series provides a forum for empirical and theoretical work on Japanese business enterprise, Japanese management practices, and the Japanese economy. Japan continues to grow as a major economic world power, and Japanese companies create products and deliver services that compete successfully with those of the best firms around the world. Much can be learned from an understanding of how this has been accomplished and how it is being sustained.

The series aims to balance empirical and theoretical work, always in search of a deeper understanding of the Japanese phenomenon. It also implicitly takes for granted that there are significant differences between Japan and other countries and that these differences are worth knowing about. The series editors expect books published in the series to present a broad range of work on social, cultural, economic, and political institutions. If, as some have predicted, the 21st Century sees the rise of Asia as the largest economic region in the world, the rest of the world needs to understand the country that is, and will continue to be, one of the major players in this region.

Editorial Board

Knowledge Works

Managing Intellectual Capital at Toshiba

W. Mark Fruin

New York   Oxford
Oxford University Press
1997

Oxford University Press

Oxford   New York
Athens   Auckland   Bangkok   Bogota   Bombay   Buenos Aires
Calcutta   Cape Town   Dar es Salaam   Delhi   Florence   Hong Kong
Istanbul   Karachi   Kuala Lumpur   Madras   Madrid   Melbourne
Mexico City   Nairobi   Paris   Singapore   Taipei   Tokyo   Toronto

and associated companies in
Berlin   Ibadan

Copyright © 1997 by Oxford University Press, Inc.

198 Madison Avenue, New York, New York 10016

Oxford is a registered trademark of Oxford University Press

Library of Congress Cataloging-in-Publication Data
Fruin, W. Mark. 1943–
    Knowledge works : managing intellectual capital at Toshiba / by Mark Fruin.
       p.   cm.
    Includes bibliographical references and index.
    ISBN 0-19-508195-1
    1. High technology industries—Japan—Case studies.   2. Tōshiba, Kabushiki
Kaisha.   3. Organizational learning—Japan—Case studies.   4. Information resources
management—Japan—Case studies.   5. Industrial management—Japan—Case
studies.   I. Title.
HC465.H53F78   1997
338.4'762'000952—dc20   96-21036

9 8 7 6 5 4 3 2 1

Printed in the United States of America
on acid-free paper

Preface

For six years, from 1986 to 1992, I spent parts of each year working in Kawasaki City, just outside of Tokyo in one of the Toshiba Corporation's major (yet physically unimpressive) factories, the Yanagicho Works. Altogether I wore Yanagicho's factory uniform for a little more than a year, working at a half-dozen different locations within the facility. The first year I was attached to the manufacturing engineering support section. The second I worked on the shop floor assembling ATM machines, photocopiers, automatic ticket issuing and validating equipment, and optical-electric disk drives. During my second summer I also lived in a Toshiba dormitory for managers near the Kamata JR (Japan Railway) station. Then I looked at work design, flow, and control in various product departments and at product planning. Later I studied quality assurance, employee training, compensation, motivation, and personnel management.

In 1990–91, under the auspices of the Fulbright Commission, the Social Science Research Council, Tokyo University, and INSEAD (the European Institute of Business Administration), I spent most of a calendar year in Japan examining the relationship of Yanagicho to other organizational units within Toshiba and to evolving economic and locational aspects of high-technology manufacturing in Japan. Write-up followed data-gathering and with the help of the Institute for International Economic Studies in Tokyo and the Institute of Asian Research at the University of British Columbia in Vancouver, I was able to find the time to make sense of what I had seen, heard about, and reflected on.

A considerable period of time has elapsed from the start to the finish of this study—perhaps too much. The wonder of discovering what I believe to be a whole new genre of high-tech production sites, Knowledge Works, the term that is also this book's title, has subsided ever so slightly over the years, especially after giving two or three dozen talks on the subject. Yet the Knowledge Works concept, the most important construct to emerge from this study, undeniably represents a new way of thinking about manufacturing organization and strategy, and it explains why "ordinary" factories in the electronics/electrical equipment industry in Japan are so "extraordinary" in their capabilities. It also augments the focal factory model that was proposed in my last book, *The Japanese Enterprise System* (Oxford University Press, 1992).

By emphasizing how new and different the Knowledge Works construct is, I risk being associated with "the gee-whiz school" of Japanese Studies, something that I most certainly don't welcome. Anytime someone finds something different about Japan, he adds to the considerable literature of "them versus us" (I say, "he," because for the most part this is a male, white Western enterprise). To be fair, however, there are also counterliteratures of more recent origin that focus on the convergence as well as divergence of modern Japan with the rest of the world, and these completely bypass the issues first raised by Lafcadio Hearn in the 19th century. But they do so at the cost of deemphasizing how different the origins of modern Japan really are.

Despite my wish not to exaggerate these differences, I do accentuate how different the Knowledge Works concept is from standard models of manufacturing organization and strategy, and I cannot overemphasize the importance of the construct for understanding the success of industrial firms such as Toshiba, Hitachi, NEC, and Mitsubishi Electric. Today, these firms and just a handful of others (Korean and a few Western examples) dominate the global electronics/electrical equipment industry like never before. I believe that there is something in the structure and function of these firms and their factory systems that explains their sustained success and industry domination.

I must add that the Yanagicho Works, the main fieldwork site for this study, is not an exemplary factory among Toshiba's Knowledge Works. The average Yanagicho worker is 39 years old. The plant itself is 60 years old. Photocopiers, Yanagicho's most important product, are not the most competitive in Japan or overseas. Yanagicho falls somewhere in the mid-range of Toshiba's factories with respect to the number of technically trained employees on-site and the value of outsourced parts and components as a percentage of production value. In other words, I have not chosen to investigate a particularly outstanding factory, and yet what I have found is remarkable and well worth further study. That is the real story of Yanagicho and of Toshiba's competitive success.

Knowledge Works is a qualitative concept, based on inductive theory-building in the field and historical, industry, and company sources. *Knowledge Works* is syncretic in method, interdisciplinary and multidisciplinary in approach. Contrary to what some disciplinary purists might claim, I strongly believe that the richness of these sources, methods, and perspectives adds rather than detracts from the study. And given that our knowledge of organizations, institutions, management, and strategy in Japan and in the rest of Asia is in its infancy, compared with what is known about these matters in the West, clearly a syncretic, inductive, and catholic sort of approach is called for.

Finally, I should mention that I have chosen to include and interlace only a selection of the materials that I have collected inside Toshiba. I could not possibly include everything, both because this book would be too long and because I am not able to make sense of everything that I have collected. Making sense of everything in the midst of everything is a formidable tast. These limitations affect not only what I write but also how I write it.

Data are not always as current as they could be, but this is related to both methodological and conceptual isues. As standard procedure for someone concerned with the *evolution* of corporate strategy and structure, I collected all kinds of data on a dozen or more trips to Japan, and I provided a draft of the manuscript to a number of top and middle executives of the Toshiba Corporation. This was to ensure the accuracy of the information in the manuscript and to allow Toshiba to comment on data that they felt might be proprietary or sensitive.

The executives said that I could have more current but less precise data and thereby satisfy my desire to look at the evolution of Knowledge Works practice. Or, I could keep the raw numbers and data that I collected and thus retain the accuracy of the information found in this study. Faced with this choice, I elected to keep the information in the study as precise and accurate as possible, believing as I do that the strength of this study is found in its field-based method and in the quality and integrity of the observations and conclusions that I have generated from the field.

Obviously it would be desirable to have not only updated numbers, and lots of them, but also comparative data for a number of different factories. Such desires underscore the frustrations of field-based research. One can look long and hard at some sites but not everywhere at once. Also, Toshiba did not hire me or engage me as a consultant, and in order to gather similar, in-depth data at other Toshiba factories, I would have had to spend several more years in the field, which for financial and personal reasons I chose not to do. However, I was able to gather some comparative data on the number and type of employees, breadth of products, marketing strategies, range of multifunction capabilities, and nature of suppliers at a number of Toshiba plants, and with this information (much of which appears in this book) I was able to satisfy at least some of my concerns as to the representativeness of what I discovered at Yanagicho.

Neither the data gathering nor the write up were easy. I don't know how many times I asked myself whether or not I was too old to be climbing yet another set of stairs shoulder to shoulder with younger men who, among other things, were being paid to climb stairs. At the end of the book, I list more than 60 formal interviews that I conducted in the course of this research, and that only begins to suggest how many times and of how many people I asked questions and sought information. And I certainly don't want to calculate how many hours I spent at my Toshiba 4400sx trying to say things clearly and well. The capacity of the project to absorb labor without approaching a marginal rate of return was surprising. It made me doubt both the utility model of labor and my writing abilities.

Debts pile up quickly in this business because family, friends, and colleagues are an endless source of good ideas. In this vein, I have to single out Paul Adler, Franco Amatori, Richard Appelbaum, Maria Arias, Mary Yoko Brannen, Alfred D. Chandler, Jr., Bala Chakravarthy, Kasra Ferdows, Martin Fransman, Nathan Fruin, Noah Wardrip Fruin, Richard L. Fruin, Jr., Martin Kenney, Jean-Pierre Lehmann, Jeffrey Liker, Leonard Lynn, Masao Nakamura, Robert C. Miller, Tom Roehl, Tom Rohlen, G. William Skinner, Gianluca Spina, Brian Talbot, Ilan Vertinsky, and Eleanor Westney, all of whom have contributed in tangible and intangible ways to the completion of this book. In particular, I want to thank Mr. Herbert J. Addison of Oxford University Press who has been extraordinarily encouraging and supportive of this project from its inception.

To my many friends at Toshiba, I am forever grateful. Without their help, I would have accomplished very little. A list of those who helped me would run to several pages, so I am able to mention here only the most outstanding men and women of Toshiba who contributed to this effort. But first, not only because he is most deserving but also because he is no longer with Toshiba, I must single out Masaharu Tanino. He was my most important contact within Toshiba and he opened many doors for me. After Mr. Tanino, the following list of persons "behind the doors" is partial at best and in no particular order of importance: Takaomi Aoki, Sanae Demizu, Hitoshi Dojima, Mitsuo Komai, Akira Kuwahara, Masaaki Inokuma, Hiroyuki Matsuda, Tomonobu Shibimiya, Hiro Shogase, Masuo Tamada, Dr. Sei'ichi Takayanagi, Hiroshi Tsutsui, Kazuichi Wada, Masanori Yamoto and Ryoji Yamoto. *Mina no koage de dekiagarimashita. Arigatoo gozaimashita.*

Finally, I owe a great deal—at some level almost everything—to my family. My parents, Richard Lawrence Fruin and Gertrude Winter Fruin, both of whom recently turned 82, encouraged me when I first went to Japan as a nineteen-year-old Stanford student. They supported what began as "a junior year abroad" and turned into a lifelong occupation and preoccupation with Japan and Asia. My brother and sisters have always been there, too, cheering on my intellectual and physical travels in Asia. My sons, Noah and Nathan, have accompanied me on many of these trips and through their eyes I have learned even more. Noah is a kindred researcher and writer, someone who never tires of examining how critical theory, factory-based fieldwork, and organization studies might come together. Indeed, my ideas on the electronic workplace metphor may be traced to our many discussions of the Internet and interactivity. Also, many of this book's illustrations and figures were inspired by Noah and refined by Michael Crumpton, his colleague at the NYU Center for Digital Multimedia Studies. Nathan, within a few visits to Yanagicho, knew more about the production processes there than months of careful observation had garnered me. Nathan—not I—is the real entrepreneur and applied scientist of the family, and his spirit enlivens many parts of this study. Nora, my six-year-old stepdaughter, is used to adults punching away at computers all day long, and she did the best that she could do in these circumstances. She stayed out of the way.

Mary Yoko Brannen, my partner and closest colleague, knew what to do and how to do it when I ran into the inevitable blinds of "writer's block." These are as much spiritual as mental in my opinion. She has a true gift of interpersonal, intercultural, and spiritual understanding, and she deserves more credit for helping me complete this book than a few lines in any preface can convey. My family, in other words, sustained me throughout this study and it is to them, once again, that I dedicate this book.

Lummi Island, Washington   W.M.F.
Autumn 1996

Contents

KNOWLEDGE
WORKS

ONE

Form, Creativity, and Competitiveness

Japan's industrial competitiveness is factory-based competitiveness, and for Toshiba, Knowledge Works are the source of its competitive, leading edge in manufacturing. Knowledge Works are focused, yet flexible and versatile, high-tech manufacturing sites that emphasize an integrated management of people, capital, and technology, and embody an abundant base of knowledge, capability, and ambition. This factory-as-fulcrum argument goes beyond Alfred Chandler's 20th-century emphasis on vertical integration, management of physical assets, and the coordination of scale and scope economies to argue that network forms of organization, workplace-centered human resource strategies, and intangible asset management constitute a new paradigm for factories and industrial firms in the 21st century.

The Knowledge Works model explored and advanced in this book is a grounded construct for managing high-tech, factory-based innovation and renewal. It is grounded in 6 years of studying the Toshiba Corporation, itself a product of 120 years of history; a full year of participant observation in the Yanagicho Works (one of Toshiba's 27 domestic factories); plus 20 years of reflection on the history of industrialization, firm organization, and competition in high-tech industries. But, above all, my ideas are grounded in friendship. Friends, on and off the line, in and out of Toshiba, opened my eyes and made me think hard about factories and industrial life.

Allow me to extend to you the same opportunity. Let me open your eyes and walk you through Yanagicho with my experience and knowledge as your guide. Yanagicho is a factory site of 113,669 square meters with floorspace of 135,264 square meters in Kawasaki City, just across the Tamagawa River to the southwest of Tokyo. Yanagicho is Toshiba's primary production facility for the manufacture of plain paper copiers (PPCs), laser beam printers (LBPs), optical disc drives, and a number of labor-saving devices such as automatic mail sorting equipment, railroad and turnpike ticketing systems,

Figure 1.1 Large-letter campaign banners inside the main gates of Yanagicho.

and automated teller machines (ATMs). Characterizing Yanagicho is difficult because so many different products are made there (13 at present); so many different production and assembly technologies—everything from high-volume automated lines to custom-order one-off lines—are deployed there; so many distinct functions—everything from R & D, design, and engineering to quality assurance, product planning, and marketing—are combined there.

After climbing the steel walkway over the busy Keihin Expressway fronting Yanagicho, the first thing to capture your eyes is how utterly ordinary and neat the buildings appear. Every horizontal surface is swept and clean, everything is in its place. There is nothing out of the ordinary, nothing to suggest that this is one of Toshiba's leading factories for making sophisticated products like high-speed, full-color, double-sided photocopiers and SuperSmart cards, credit cards that put the power of computers at your fingertips. (See chapter 5 for a description of the development of this pathbreaking product.)

Clearing the front guard station, where employees flash their badges and visitors register their intentions, a three-story, 50-year-old (but freshly painted) administrative building sits on the right. A border of greenery runs from a small Shinto shrine on one end to some large-letter banners on the other. The banners read "Creating New Values," "Safety First," and "New Spirit Campaign" (the first two banners are in Japanese; the last in English). The letters can be read easily from a 100 meters away. The banners frame a busy, well-

Figure 1.2 The traditional face of manufacturing: row upon row of ferroconcrete blocks.

marked thoroughfare where people, trucks, pallet lifts, and visitors stream from behind the administrative complex to a line of imposing, three- and four-story, well-proportioned buildings, sweeping left and right (see Figures 1.1 and 1.2) .

The hurry-scurry and muffled noise around the inner buildings suggest that now we're in a place where something industrial is happening. The neatly signed roadway and ribbons of greenery pull one along, and turning left between two towers one spies two more four-story blocks. At the end of the row lies a sports complex where baseball and basketball teams vie for industrial league recognition; to the right stands a six-story, steel and glass skyscraper that contrasts sharply with the smaller buildings crowding it. That's the *gijutsuto* (literally technology tower), or so-called Power Tower, that is increasingly defining the nature of industrial work. Employees there are more likely to be operating CAD/CAM stations than NC (numerically controlled) drill presses (as shown in figures 1.3 and 1.4).

While the buildings are neat, well painted, and well cared for, none of them are especially noteworthy except, perhaps, for the Power Tower. Until one goes inside, that is. Inside, one is struck by the brightness, neatness, and orderliness of what might otherwise be a messy, dark, dirty, and disordered workplace. After all, tons of steel alloy, aluminum, and plastic parts and components are turned, shaved, cut, pressed, drilled, and kitted here, and then assembled into thousands of laser beam printers, compressors, photocopiers, ATMs, utility meters, and railroad ticket taking and validating ma-

Figure 1.3   The six-story-high, football-
field-long Power Tower (*gijutsuto*).

Figure 1.4   Manufacturing's new face: computer-aided
design, development, and manufacture.

chines. Where is the helter-skelter disarray and clutter that are routinely associate with high-volume factory work?

Not swept under the rug. Hardwood oak floors in the older buildings gleam with wax polish and years of use. Wall surfaces and vertical columns bristle with posters, signs, diagrams, and flow charts. Floors are clearly marked with white, yellow, and red lines. There is order, orderliness, and organization everywhere. One hesitates to say that everything is in its place because how can anyone know where everything should be, but in fact order, orderliness, and organization are certainly the impressions given along the transfer lines pumping out photocopiers and laser beam printers (see Figures 1.5 and 1.6).

The lines run full and steady, from 8 A.M. to 5:30 P.M. on a normal workday. When glitches arise, workers, engineers, and if necessary, managers huddle in face-to-face, hands-on problem solving (see Figure 1.7). Noise is the only giveaway that this is an industrial workplace. A regular thump-thump and eeerh-eeerh of metal on metal punctuates the shop floor. The intensity and wavelength of the noises, and the number and kinds of workers depend on where we are in the complex. The production line for SuperSmart cards, for example, is shielded behind thick glass walls where workers are sheathed, head to toe, in smocks, caps, and rubber gloves like those worn by doctors in surgery. Because manufacturing yield is directly related to cleanliness, and thus to reliability and profitability, on the smart card line cleanliness is really "next to godliness".

Figure 1.5 Order, orderliness, and organization along the photocopier line.

Figure 1.6   Passed inspection and ready for shipping at
the end of the line.

Figure 1.7   Face-to-face problem-solving off the line.

Assembly on the meter equipment line tells another story. Here
we find the bangs and booms of metal beating on metal, and the
heat, steam, odor, and noise of what most of us think of when we
think "factories." The meter line is the factory's oldest, reflecting the
traditional metal bending and cutting technologies of early-20th-
century factory work. Many at Yanagicho say that they would be
happy to be rid of the line and that its traditional metal bending

processes are out-of-sync with the rest of the place, but they also acknowledge Yanagicho's responsibility to provide employment for meter line workers and for the dozens of meter parts and component suppliers nearby. Stability of employment, even at the cost of some profits, is a big part of Yanagicho's story.

Stable and, hopefully, satisfying employment does not go hand-in-hand with keeping old production lines running, however. In today's demanding markets, innovation is the key to stable employment because consumers always want the latest and greatest, especially if they can have it at fairly constant costs. So, improvisation and innovation are the real jobs of workers at Yanagicho. Along with Power Tower engineers and technicians, Yanagicho workers are charged with the mission of making things better, quicker, and cheaper.

Yanagicho has a high ratio of managers, engineers, and technical employees to direct workers, attesting to the diversity of its product lines and product support functions. Only one-third (36.4 percent) of all regular employees are production workers. The rest are concerned with design, development, engineering, sales, inventory, quality, as well as accounting, information systems, personnel, and so on. In 1991, 505 employees were attached to Yanagicho as researchers and designers in the Power Tower, raising the percent of knowledge workers on site to a remarkable 85 percent (see Figure 1.8). And these totals do not include a sizeable number of non-Toshiba engineers and technicians from affiliates and suppliers, such as the 280 employees of Toshiba Intelligent Technologies, Inc., who were also working at Yanagicho in 1991.[1] With more than 4,000 workers on site, more than a dozen product lines, and everything from R & D to shipping locally available, Yanagicho is both a smorgasbord and a cornucopia of people, products, processes, and engineering systems, as illustrated in the Appendix, where a department and section organization chart for the factory may be found.

Yet whether one makes smart cards, designs photocopiers, assembles meters, or labors anywhere at Yanagicho, the job is the same: work with your hands, head, and heart. Work with brawn, brains, enthusiasm; work together because making better products benefits you, your workmates, your section, department, factory, and company—in that order. Such commitment and conviction are manifest everywhere, but perhaps they are most apparent in the eyes and expressions of Yanagicho's industrial trainees.

They receive a year or two of advanced technical training in machine tool operations, software programming, and production operations at Toshiba's expense after graduating from high school. Before the 1970s, industrial training programs like this one were high

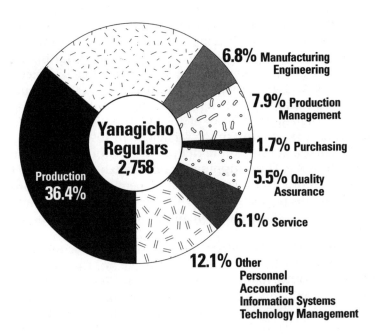

| | Info Systems Division 128 |
| Info Systems R&D 194 |
| Software R&D 150 |
| Central R&D 27 |

	Regular Employees			Resident Employees	Totals
	Indirect Employees	Direct Employees	Totals		
Information Systems	1,728	676	2,404	322	2,726 (84%)
Electrical Equiptment	101	131	232	6	238 (7%)
Consumer Electronics	35	87	122	0	122 (4%)
Other Areas	0	0	0	177	177 (5%)
Totals	1864	894	2,758	505	3,623 (100%)

Yanagicho Regulars 2,758

Production 36.4%

6.8% Manufacturing Engineering

7.9% Production Management

1.7% Purchasing

5.5% Quality Assurance

6.1% Service

12.1% Other
Personnel
Accounting
Information Systems
Technology Management

Figure 1.8 Knowledge workers at Yanagicho Works.

Figure 1.9 Yanagicho's trainees seem happy in their work.

school equivalency programs, and so workers were doubly indebted to employers both for a job and a high school education. Today, however, everyone entering Toshiba already has a high school diploma, and so the training focuses narrowly on machine tool skills, assembly techniques, and some operations management. Trainees choose between staying on as regular employees or working elsewhere at the end of their training (see Figure 1.9). Not so surprisingly, most stay on.

Seeing their enthusiasm and comradeship is an eye-opening experience, especially in a world where industrial work has lost much of its appeal and occupational calling. Yet Knowledge Works are not production sites of some imagined time and space; they are present in ever growing numbers in Japan, and one hopes they will spread overseas as a means to organize and manage the industries of tomorrow.

Today, Not Tomorrow

The enthusiasm, camaraderie, and calling of Yanagicho's workers clash with our preconceived notions of factory work. For the past century, science fiction authors such as Jules Verne, H. G. Wells, Philip K. Dick, and Ray Bradbury have posed seemingly farsighted visions of the future of manufacturing. They conjure an ill-omened, high-tech future where technology and capital substitute for labor, and where workers, the few or many of that species who survive, are highly constrained in the production process.[2]

The business press confirms some of that disquieting vision. Technology and capital appear to be replacing and displacing labor on an increasing scale. Major debt crises in Canada and Mexico, lay-offs in America's Midwest, Northeast, Northwest, and Southwest regions, and downsizing in motor vehicle, computer, capital equipment, aircraft and aerospace industries are rife. The news from Western Europe is also discouraging: high national debt and unemployment in advanced industrial economies accompanied by new investment in less developed parts of the world, such as Eastern Europe, Southeast Asia, and South America, with more accommodating workers, fewer welfare entitlements, lower wages and taxes.

But the story is quite different in Japan. Japan is one of the most successful industrial economies of this century, sustaining high rates of growth and productivity gain for the past half century, not to mention the previous 40 to 50 years of better than average performance. Yet in Japan's large firms, technology and capital do not readily substitute for labor; during the decades of most rapid economic progress, the 1950s to 1980s, and even in the midst of the prolonged recession of the early to mid-1990s, firms did not lay off employees except in the most extreme circumstances. Part of the reason for this is that most employees of large firms—not just the workers—are union members. So, you are paid according to years of seniority among other things. Formerly, seniority weighed more in wage calculations; today it counts for less. The seniority component of wage increases in 515 firms was 38.5 percent, according to recent research.[3] Hence, there is a rough correspondence between years of employment and size of one's paycheck, because, in Japan, both are related to growth in labor factor productivity.

In the last few years, in the depths of the worst recession since the 1950s, there have been a number of well-publicized layoffs and downsizings in Japan. But the 800 to 1,500 workers who lost their jobs with the likes of Mitsubishi Heavy Industry, Toyota Motor, and Ishikawajima Harima pale to insignificance in comparison to the tens of thousands laid off by AT&T, General Electric, General Motors, Boeing, McDonnell Douglas, General Dynamics, United Technologies, and so on. Reality and cultural expectation still favor long-term employment, seniority-weighted promotion, and union membership for all regular employees of large firms in Japan.

Unable to dismiss and lay off or to neglect and underpay, except in the most extreme instances, large industrials have moved in another way, where regular employees are considered to be irreplaceable firm assets. Hence, during periods of the most rapid economic growth, the 1950s to 1980s, and even the slowdown thereafter, Japan's industrial firms are noteworthy for their balanced and prudent strategies that pivot on committed human resources, capital-intensive technologies, and factory-based organizational capabilities. Such strategies are nowhere more evident than in the electronics/electrical equipment industry. Toshiba competes successfully in this industry–one where the interplay of capital, technology, and labor is exceedingly intense, and one in which Toshiba faces the same price and cost pressures as General Electric, Philips, or Hyundai but without their layoffs, labor strikes, and radical restructuring.

Between 1990 and 1993, Japan's major industrials realized significant gains in world market shares against all major Western competitors in computer/office equipment, nonelectrical machinery, autos, chemicals, electrical equipment, textiles, tires, iron and steel.[4] They did so in large measure because of production facilities like Toshiba's Knowledge Works, where the point is high-value-added production, not just production. High-value-added production flows from the integrated and interdependent management of people, technology, and capital, bestowing distinct and growing competitive advantages on Toshiba, even as the electronics/electrical equipment industry globalizes and becomes more competitive.

High-Value-Added Production, Not Simply Production

"The $1=80 Yen Factory," declared the November 19, 1994, front cover of *Nikkei Business*, the authoritative weekly of the Nikkei publishing group. At that exchange rate, the suggestion was that Japan's factories would have a hard time competing. Barely a month later, the front page of *The New York Times* business section proclaimed "Stunning Changes in Japan's Economy".[5] Both *Nikkei Business* and *The Times* were arguing that Japan's economy was undergoing major, even fundamental, restructuring. And less than three months later, the exchange rate indeed slipped to an all-time, postwar low of less than 80 yen per U.S. dollar.

"Making Things that Challenge a Hollowing-Out of Manufacturing," continued the *Nikkei Business* story. "Hollowing-Out" (*kudouka*) is the Japanese term for losing jobs to lower cost and overseas workers and facilities. "Challenging *kudouka*" suggests that Japanese manufacturers are taking seriously the rising value of the yen and growing foreign competition. *The New York Times* declared that they were doing so by moving more production and assembly to low-cost, offshore sites in Asia and the West, by consuming more at home, and by moving into higher-value-added, information-based industries and service activities.

The changes pinpointed by *The Times* are not new. The yen's high value forces Japan's firms to do what they have already been doing, only more so, again and again. For nearly a decade, beginning with the oil crises of the 1970s, firms have been responding to unfavorable exchange rates, commodity price fluctuations, production overcapacity, and trade-related political pressures by innovating managerially, organizationally, and technically.

If innovations create more value than costs at home, and this is the point, firms *will not be stampeded* overseas with a grim prospect of ever greater unemployment and underemployment at home. Instead, by increasing productivity and sustaining high levels of new product and business development, *koudoka* will be arrested and firms will not be caught between the developed world on one hand and the developing world on the other. Firms can choose to go overseas for any number of good reasons, and they are likely to be even more successful, based on what they have already learned at home. Factory-based experimentation with and refinement of routines, skill sets, and practices are among the most useful successful things learned.

Knowledge Works argues that a strategy of constant, across-the-board innovation and renewal at the factory level of organization has enabled Japan's most successful industrial firms to create more value than costs in spite of an enormous 80 percent revaluation of the yen since 1980. In doing so, these firms have forged an industrial workplace quite unlike anything the world has ever seen before. These are factories where most workers have the equivalent of a junior college degree or better; where new products can be brought on stream with a minimum of additional cost; where innovation is anticipated, not exceptional. These are factories that embody substantial domains of knowledge and expertness, and where *organizational capabilities* are more than adequate to cope with uncertainty and rising costs.

Such factories are Knowledge Works, my term for what look like 21st-century workplaces at Toshiba. Only Knowledge Works are already here, and they are transforming how firms are organized and what advanced manufacturing entails in Japan. Otherwise, it is impossible to explain the success of Toshiba in today's highly competitive electronics/electrical equipment industry.

Long Live Adam Smith?

Knowledge Works suggest that in several important ways, Adam Smith's purely market-based theory of economics was right for his day but for wrong for ours. First, except for church and state, there were no large commercial or industrial organizations in 18th-century Scotland. No General Motors, no General Electric, no IBM or Royal Dutch Shell; neither a national distiller, say Glenlivet, nor a national voluntary organization, like the International Red Cross or the Salvation Army. Even church and state, harbingers of the organizational revolution to come, were loosely centralized, operationally weak kneed, and geographically far-flung. For such reasons, parish parsons and local officials operated pretty much on their own.

Today's GM, IBM, or International Red Cross are even more far-flung yet anything but loosely centralized and operationally weak kneed. Hence Adam Smith's 18th-century concern with markets and market regulatory mechanisms made perfectly good sense for his day but not for ours. Today large business organizations mediate, intermediate, and some would say, manage markets in every advanced industrial economy around the globe.[6] Second, production and distribution technologies were elementary in Smith's day, and ridiculously primitive by our standards. Oxen hauled goods at one or two miles an hour; shoemakers turned out a pair of sturdy boots after three or four weeks of dawn-to-dusk labor. A baker's dozen was a fairly large unit of transaction. Scale, scope, and speed of operations were unimportant, and hence, so was management. Management, in the modern sense of organizational authority, coordination, and control, was beyond Smith's learning and appreciation.

Third, Smith's sparsely settled Scotland was not crowded with grey-flanneled managers like contemporary New York, Tokyo, Seoul, and London. Measly, low levels of resource exploitation and conversion demanded neither row upon row of skilled and semiskilled workers nor serried ranks of secretaries, clerks, and professional managers. Large, complex organizations and the concerns of modern business—the separation of ownership and control, bureaucratic authority, professional management, entrepreneurship and intrapreneurship—were not at all Smith's concerns. Smith could not have possibly pictured the modern corporation, its market-organizing mechanisms, its strategic information systems, its echelons of highly educated, disciplined, and paid managers and workers.

The organizational and managerial differences between Smith's day and ours are truly remarkable. His world was not tightly interconnected and organized, spatially and temporally. Ours is. He was not at all concerned with management; we are. Labor moved to where capital was available in his day; capital moves freely in search of inexpensive labor in ours. Large, complex, far-flung, information-intensive organizations were not the viscera of commercial, intellectual, and political life for Smith. They are for us. Such organizations—the distributed brains of advanced industrial civilization—are spotlighted in this study of how Japan's firms manage and organize high-tech industries that are information intensive and human centered.

In spite of two centuries' accumulation of knowledge since Adam Smith, economic theory still has difficulty explaining why certain organizations perform better than others and why Japan's industrials are world-class competitors. Knowledge intensity and human-centeredness, both hard to specify and measure, are my answers. In tightly scheduled, complex production routines human ingenuity and creativity are paramount, and nothing really substitutes for highly trained, motivated, and well-managed workers. In other words, Japan's industrial firms have pushed the organizational metaphor farther, faster, and "better" than the rest of us. They are better organized, better managed, and better performing (notwithstanding the successes of Hewlett Packard, Intel, and Boeing). Organizational and managerial differences are behind what is "better" in that the growing intellectual content of work in Japan demands better employees, better-managed organizations, and better managers. Smarter is better in an age of smart products, smart machines, smart competitors, and smart organizations.

"Smarter" is found in socially, technically, and administratively advanced manufacturing sites where technical virtuosity, learning, and organizational renewal impart enormous flexibility and resourcefulness. Note the emphasis on sites. This is where I part company with Peter Drucker's writings on knowledge workers or Robert Reich's symbolic analysts.[7] Knowledge workers and symbolic analysts cannot exist apart from industrial workplaces because advanced manufacturing techniques depend on advanced manufacturing organizations. *Productivity and efficiency depend on how work and workers are organized and managed at specific factory sites.*

In other words, there are real limits to the electronic cottage metaphor. Electronic togetherness does not substitute for human togetherness when work processes are organizationally complex, varied, and interdependent, and when problem solving depends on implicit know-how and face-to-face cooperation. "Workplace smartness" is highly contextualized, so organizational and cultural differences, even within the same firm, are all-important in transforming industrial work from Adam Smith's day to ours.

Transforming Manufacturing Strategy

The phrase *Knowledge Works* implies several things. First, it suggests industrial organizations in which the production of knowledge and know-how are more important than the manufacture of things. Intangible rather than tangible goods carry the day.

Second, it describes production facilities, routinely undifferentiated as "factories," where most everything changes, often at remarkably rapid rates. Product and process engineering innovations are rife. Workers and managers are constantly interpreting and reinterpreting how they work and, to a lesser extent, why they work. Nearly everything is changing, nearly all the time. Hence, Knowledge Works employees are not machine appendages, supplying simple, monotonous, repetitive labor services. In contrast to Marx and Engels, *Knowledge Works* argues that a factory's physical endowment, its plant and equipment, its tools, jigs, and numerically controlled machine tools, are really appendages of knowledge-seeking and knowledge-exploiting workers. By dint of their intelligence, social organization, and personality, employees run and regulate a plant's everyday capacity and productive efficiency.[8]

Third, even while knowledge workers remain tied to machine-pacing and factory scheduling, they are engaged and integrated with groups and teams that are the fulcrums of industrial design, flow, and organization. Local knowledge decides issues of output, reliability, and quality. So, Knowledge Works grow and prosper not by a physical and spatial division of labor, as told by Adam Smith, but by a physical, spatial, and functional *integration of labor and information*. By tying these desirable traits to membership, promotion, and reward in multiproduct, multifunction plants, property rights, or the normal expectations of profiting from what one knows and does, are preserved. Commitment, teamwork, know-how, and profit constitute a single thread of organizational membership.

Knowledge Works focus on developing new and better products, advancing manufacturing technology, and increasing value through lowering costs, improving quality, and speeding the development process. These are goals in process, and they are intangible. Western management is mostly concerned with efficient use of tangible resources, such as buildings, labs, patents, equipment, fleets of automobiles, and the like. These are things that can be counted, routinely depreciated, and easily valued; balance sheets where people are a residual resource, not a core one.

But when people are a core resource, that is, when people are in a very real sense the company, maximizing gains from intangible assets is the heart of business strategy because only people generate knowledge and know-how.[9] Because of this, Knowledge Works may not readily promote the general welfare. The wisdom and wealth of Knowledge Works are increasingly tied to specific factory sites and organizational relations. Hence, corporate (and factory) advantages may not translate one-to-one into country-competitive advantages. Toshiba's success is not necessarily Japan's success. Company policies, by logical extension, do not equally favor all populations and economies. So much for trade and investment theory based solely on the comparative advantage of nations. In today's globalized arena of competition, with capital and technical resources flowing freely between nations, it is the percentage of high-value-added employment, as Robert Reich, Lester Thurow, and others argue, or the Knowledge Works quotient in my words, that really constitutes the competitive advantage of nations.[10]

The Problem Set

Issues of productivity, product-based competition, industrial innovation, and the intellectual content of work are straightforward enough. In a technology-driven world, more R & D is better. More includes notions of quantity and quality, and these lead naturally to considerations as well of how and why. More R & D of the *right* sort, conducted *effectively* and managed *efficiently*, will always triumph over less of these. The rule applies equally well to public and private undertakings, so that the sum of R & D, conducted effectively and managed efficiently, is a rough measure of technical competitiveness.

While the issues appear simple enough, actually they are not. According to National Science Foundation data (and, for that matter, everyone else's), output per worker and total productivity increase in manufacturing have been growing much more rapidly in Japan than in the United States for at least 30 years.[11] This has been true in spite of relatively larger shares of GNP, GDP, and industrial revenues devoted to R & D in the United States. But quantity alone is not the determining issue, although quantity may be an issue if vast amounts of money spent injudiciously skew R & D systems toward the wrong goals and wrong directions.[12] Beyond certain levels of expenditure, however, R & D effectiveness is more an issue of quality, that is, the quality of knowledge generation, acquisition, and application.

This study considers issues of quantity to be largely in the kin of macroeconomics and outside the scope of this work. Instead, microorganizational issues are the main foci of *Knowledge Works,* and so qualitative issues weigh heavily in this book. However, at some level, it becomes difficult to separate the two, especially when and if institutional patterns are repeated across industries and may come to represent, as a consequence, countrywide practices.[13] While this study concentrates on the electrical machinery/electronics industry, similar institutional patterns of integrating and localizing R & D management at the level of the factory may be observed in other R & D intensive industries where time to market competition reigns.[14] If the patterns are repeated broadly enough, the Knowledge

Works model approaches a national pattern of R & D organization and management. In this way, macroeconomic and microeconomic issues become inextricably bound.

At the macroeconomic level, to summarize the argument in the economics literature, the superior performance of the Japanese economy and the relative decline of the American economy have been related to differences in savings and investment rates,[15] changes in the aggregate stock of R & D,[16] different degrees of efficiency in the transfer of R & D to production,[17] an increase in R & D intensity in Japan,[18] different styles R & D team management,[19] changes in the proportion of R & D expenditures going to basic research,[20] and learning and price effects.[21]

Surprisingly enough, little effort has been spent relating macroeconomic data to microeconomic or organizational issues. The productivity and product development debates have been largely disembodied from people and from the factories in which they work. Herein lies the rub, if structure and manufacturing strategy make a difference. There has *not* been a major, new, fieldwork-based study of a Japanese factory for more than 20 years, that is, since Ronald Dore published *British Factory-Japanese Factory* in 1973.[22] A lot has happened in Japan since the first petroleum oil crisis.

There have been factory studies at the aggregate level, however. In 1979 Robert E. Cole brought out *Work, Mobility, and Participation*, a macrosociological, comparative study of workers in Detroit and Yokohama.[23] And, in his more recent book, *Strategies for Learning*, Cole locates the reasons for the success of quality control (QC) circles to differences in the willingness and ability of firms in the United States, Sweden, and Japan to build *nationwide* organizations in support of QC activities.[24] Even more recently, Lincoln and Kalleberg's comparative study of manufacturing facilities, *Culture, Control, and Commitment*, like Cole's books, also ponders the big picture, based on survey instruments that take entire factories as units of analysis. This approach, a standard one in cross-cultural studies, tends to minimize not only the internal differences within factories but also the differences of meaning that employees attach to work and to the social and personal relations that develop there.[25] Workplace culture and site-specific management disappear in aggregated data sets and in research designs where Japanese and American organizations are taken as functional equivalents.[26]

So while manufacturing has been in the spotlight, factories, where manufacturing takes place, have not. The negligence is especially apparent in the case of Knowledge Works where output per worker and total productivity increase have been most pronounced. But this book argues that:

1. Factories are an effective, generic, and strategic way to organize in R & D intensive, technologically turbulent industries.
2. There are important differences in structure, process, function, and meaning of industrial work at the level of factories
   in Japan.
3. These result from strategic decisions taken and implemented
   by managers.
4. They powerfully affect work output, content, satisfaction, and
   significance.

These differences emerge from notions of how to organize department and subdepartment functions, how to encourage organizational learning, cross-functional collaboration, product and process
innovation, and most importantly, how to relate organizational
subunits to one another and to whole factories through such techniques as organizational campaigning. In short, microlevel differences, as much social and political as economic, best explain the
sustained high levels of output and productivity found in Knowledge Works. So, while Toshiba and, more widely, high-tech Japan
may be a model for *Knowledge Works,* Japan is not the market for this
concept and its description in this book. Countries such as Canada,
the United States, South Korea, and any other country, advanced or
advancing, that hopes to make the transition to the knowledge-
intensive industries of tomorrow are the market.

The Solution Set

The basic premise of modern production theory is focus, and so
Yanagicho's cornucopia of 13 products is a theoretical anomaly. One
can choose among efficiency, quality, flexibility, and product variety, but one cannot have all of them, at least not simultaneously.
Yanagicho sidesteps the choice by producing more than a dozen different products and hundreds of different models *simultaneously.*
Focus is pursued within product departments, yielding product variety, quality, dependability, and strategic flexibility. This approach
distinguishes factories like Yanagicho from the so-called plants-
within-plants (PWP) model of manufacturing, while encouraging
admirable economies of scale and scope, and embracing high organizational learning with low transaction costs.[27]

Economies of speed are a shorthand for these other economies.
Delivering the right product, in the right amount, at the right time,
and in the right place, assumes that issues of scale, scope, learning,
and governance have been properly taken care of. In this way, time-
to-market performance may be taken as a summary measure of enterprise and factory performance. It is speed and the intertwining of

speed with increasing returns to scale and scope, low transaction costs, and high organizational learning that best explain Yanagicho's success. Structural and economic variables, moreover, do not operate independently of the social and political conditions that prevail in the workplace.[28] An accurate rendering of the reasons for Yanagicho's success must take all of these factors into account, singly and interactively.

The Long March to Knowledge Works

Yanagicho's march to Knowledge Works status began 30 years ago. For a historical account of how this happened, see my earlier study, *The Japanese Enterprise System*.[29] In the late 1960s, Yanagicho was already producing electromechanical typewriters, mail-sorting equipment with optical character recognition (OCR) capabilities, and lots and lots of different kinds of desktop and handheld calculators. Refrigerators, freezers, and food display cabinets were made at the time but they shared little in common, except some precision machining requirements, with information and office products such as typewriters, sorters, and calculators. Toshiba chose to move refrigeration products to its Nagoya Works and to concentrate photocopier efforts at Yanagicho.

The intense product and process development activities that were concentrated on the manufacture of photocopiers, automatic mail sorting and railroad ticket equipment, and other precision, paper-handling, labor-saving devices at Yanagicho during the late 60s and 70s transformed the factory into a Knowledge Works. So Yanagicho was not created or established as a Knowledge Works; it became a Knowledge Works by combining several functions: large-scale manufacturing experience (mostly for calculators and household appliances) with intense design, development, and engineering activity (mostly for automatic mail sorting equipment and photocopiers) and with an overarching concern for quality, reliability, versatility, and time-to-market speed.

Organizational capabilities on the order of Knowledge Works were and are a longtime coming, and they come at high risk and high cost. Markets for high-tech products are characteristically volatile, often treacherously so, and decisions to concentrate design, development, and engineering resources for the manufacture of product lines are costly. Once such decisions are taken, however, emphases change; leveraging and enhancing capabilities become the key concerns. Yet how much easier to have built capabilities during the scale-oriented manufacturing of the 1960s and 1970s than to take the plunge during the 1980s and 1990s! By the 1990s, given that

75–80 percent of Toshiba's 27 domestic factories had adopted the Knowledge Works architecture, notions of how to balance scale, scope, and learning economies within the context of firm organization, country demand, and global markets have been realized at Toshiba.

The Extraordinary "Ordinary" Factory

As surprising as the technical and organizational virtuosity of Knowledge Works is the equanimity with which the Japanese view them. The term "knowledge works" is a translation of the term *kaihatsu kojo*, which is Toshiba's way of distinguishing this form of factory organization from mass production facilities. A direct translation of *kaihatsu kojo* would yield "development factory" or something equivalent. I use "knowledge works" instead because the connotations are richer, the scope broader.

Knowledge Works are ordinary organizations with extraordinary capabilities. They are ordinary because that is how the Japanese see them. It took several weeks of observation and interviews at Yanagicho in 1986 before I picked up the casual way in which factory employees referred to themselves as, "*kaihatsu kojo* workers." For them, the designation meant that their mission was not simply to make things but to design, develop, and market them as well. There are two kinds of factories, mass production and development their missions are different but their coexistence is taken for granted.

Yanagicho should not in the least be taken for granted. It is a much more relevant and important model of factory organization and management than, for example, Nissan's highly acclaimed, robot-welding factory in Zama that, by the way, closed down in 1994. Nissan, GM, or any other manufacturer with ready cash can build a state-of-the-art, robot- and computer-filled plant, witness GM's attempts at building a factory of the future. But the real challenge is to upgrade, rebuild, and nurture the *existing* stock of plant and equipment without abandoning or replacing the workers who labor there.

When a Japanese manufacturer moves from Yokohama to Michigan or Normandy, unwanted plant, equipment, products, and people are not shed in the process. With little or no experience in shutting down facilities and laying off workers, and with people their most valued asset, the lion's share of large firms stay open, invest, upgrade, and renew their internal resources through training, small group activities, and organizational campaigning. And, typically, they continue overseas with policies and activities that were really designed for home, thereby making them even more resilient when modified according to local custom. Indeed, the policies and activities that are carried afar and purposely re-created overseas suggest what could be considered the core of Japanese-style management.[30]

Yanagicho's buildings are old; the majority were rebuilt in the 1950s after World War II. Its people are old: 39.6 years being the average age for all employees, and higher yet, 41 plus, for production line workers. Yanagicho has a history, an impressive half-century history of committed people, making things, structuring and interpreting the process of work. Nevertheless, in spite of old buildings, old people, and a traditional spirit of enterprise, Yanagicho is on the cutting edge of contemporary high-tech manufacturing in optical-electric, information technology (IT). Yanagicho is a product of history, not hampered by or held back by history.

Notwithstanding the equanimity with which Knowledge Works are viewed in Japan, their advantages are quite extraordinary. They may be highlighted in four ways: product and process innovation, speed and cost of product development, productivity, and employee commitment. (Each of these warrants further examination in the body of this study.) Together, they suggest that Knowledge Works are *a social and organizational innovation of enormous material significance.*

General Perspectives

Knowledge Works focuses on Yanagicho, of course, but more generally on manufacturing strategies in high-technology industries and on factory-based reasons for the remarkable postwar growth in Japan's output and productivity. Knowledge Works are not isolated and infrequent phenomena. A majority of Toshiba's domestic factories fall into the category, and an even higher percentage of Hitachi's robust factories could be similarly categorized. Insiders agree that all major electronics/ electrical equipment manufacturers in Japan employ some version of the Knowledge Works model, while high-end motor vehicle makers, precision instrument producers, fine chemical and pharmaceutical firms, and nonelectric machinery makers have their own industry-specific versions of the architecture.[31] As a structure and strategy for learning and embodying knowledge, the model is well understood and practiced.

Large industrial firms have been systematically differentiating facilities and functions to achieve and sustain strategic advantages in product design, market penetration, and product life-cycle acceleration. These advantages, developed first at home but now spreading regionally and globally, lie at the core of Japan's success in manufacturing and technical innovation. The creation, organization, and management of these advantages define the Knowledge Works architecture and they increasingly differentiate Japanese, North American, and European approaches to productivity and competition.[32]

Knowledge Works are a force for a new age, a *postproduction* age, of intellectual capitalism. Instead of making things, a production problem pure and simple, making the right things, in the right amounts, at the right times and prices, is the postproduction problem. And because nothing stands still, making *righter* things, in *righter* amounts, at *righter* prices and times, must be the goal. Postproduction, or intellectual capitalism, presumes information processing abilities of a high order, on-site differentiation and integration of functions, a customer-is-always-right point of view, and quite emphatically, an environmentally conscious mode of operations. In order to make no more than necessary, today's politically correct and sustainable development viewpoint, everything has to be organizationally and informationally integrated.

Knowledge Works are more than sites or zones of flexible specialization, another new manufacturing paradigm. In this model, flexibility resides in the capacity to mobilize many different kinds of resources on a regional basis. The need for flexible specialization arises because large industrial firms do not have the organizational capacity to combine scale and scope economies effectively, resulting in a chronic oversupply of "me-too" products. But, in fact, the individual parts of the "flex-spec" system are neither as differentiated nor as integrated as in the Knowledge Works model.

Higher levels of differentiation and integration are realized in Knowledge Works because they internalize higher levels of hierarchical coordination and control, not to mention strategic leadership. A unified strategy, structure, and culture culminating in a single firm's strategic vision holds everything together. The flex-spec model does not measure up to the Knowledge Works standard because locality ties limit the range of investment choices, strategic competence-building, and hierarchical response.[33]

Economic Perspectives

Since there is an abundance, really an overabundance, of information—market information, product information, information about changing consumer preferences, exchange rate fluctuations, technology trends, and the activities of rival firms—the problem is creating something of value amidst information abundance. When information is a commodity, in other words, access to information imparts no strategic advantages. Herein lies the value of an accumulated and refined product-making memory. Knowledge Works inculcate experience with respect to certain engineering processes as well as the making and marketing of certain products. That repetitive experience embodies the wisdom of a factory's workers and managers, a wisdom of how to make certain products quickly, economically, and well. Such wisdom is specific to certain sites and social relationships, and it is decidedly not free.

So while information may be free or nearly so, organizational learning and knowledge are not. They are costly because they have been processed in some way and made useful in some fashion. Organizational processing is the most expensive part of the information-to-knowledge transformation; Knowledge Works economize on these costs and capitalize on their applications. Such comparative organizational advantages are the source of Toshiba's competitive advantages in manufacturing.

Knowledge is a fourth factor of production, one that is unlike land, labor, and capital, because its value is specific to particular uses. In other words, knowledge, really useful knowledge with specific organizational applications and purposes, has relatively few buyers and sellers. Since useful knowledge is sticky in any number of ways, it is expensive to restore to more generic applications, unlike land that can be bulldozed and reused, labor retrained, or capital redeployed. Obviously, there are conversion costs for transforming traditional factors of production, but the costs of converting knowledge back into more general information are more prohibitive. Further, unlike the other factors of production, the value of useful knowledge is often quite perishable; alternative uses are generally few and far between. Old knowledge, in other words, loses its value very quickly.

In effect, the utility of knowledge is directly proportional to its cost, and cost has little relationship with value. A completely idiosyncratic or highly asset specific investment is without value when it is employed for alternative uses. So while knowledge can have multiple uses, once applied, its value is hard to recover and not equal in all applications. The more specific an application, the greater the cost and the more problematic a conversion to alternative uses. At the factory level of organization, knowledge and capability—the bases of manufacturing strategy—are quite specific, and given the accumulated costs of specific product and process knowhow, the likelihood of finding alternative uses at reasonable rates of return are low.

It will pay to convert knowledge into alternative uses only when conversion and transaction costs are minimal. Such costs are low when product/process know-how and accompanying design and development requirements are reasonably similar. In short, the likelihood of finding multiple uses for knowledge hinges entirely on conversion and transaction costs. Knowledge Works save on these costs by creating a knowledge base of specific technologies, skills, and relations where embodied organizational capabilities minimize conversion and transaction costs.[34]

Cultural Perspectives

The form of Knowledge Works has cultural significance as well. There are important national differences with regard to the way that discovery, creativity, art, and imitation are expressed in work. In Japan and more generally among the nations of East Asia that share a common cultural heritage based on Buddhism, Confucianism, and elements of philosophical naturalism expressed in Taoism, shamanism, and Shintoism, energy, vitality, and creativity are expressed through *knowledge of forms*.

But all form-specific experiences are not equally valuable and, for this reason, experiences with specific forms are critical. These forms, or *kata* in Japanese (also *katachi*), are the very essence of knowledge because through experiences with particular forms, the doors of discovery and, ultimately, of creativity are unlocked. Forms formalize particular notions and actions that are thought to be especially useful, powerful, or effective. Some forms, therefore, may be far more puissant than others since they encode social discipline, specific knowledge, and experience. Hence, true believers make major distinctions out of relatively minor differences, say, between the *ura-senke* and *omote-senke* ways of doing the tea ceremony.

Outwardly tense and rigid, forms are really dynamic and personal because they embody the history of an artistic and aesthetic *michi*, a way or approach to life. Forms predispose the methods by which we learn, how we play, experiment, and experience.[35] So, minor differences in form may not be so minor after all, and thus the importance of the teachings and teachers of form. While walking in Kanda, an older "downtown Tokyo" neighborhood, in mid-August 1991, I came across an advertisement for Buddhist religious objects, like altars, rosaries, and such, that read:

> Kokoro wa katachi o motomeru
> Katachi wa kokoro o susumeru
> [The spirit (soul) looks for form.
> Form advances the spirit.]

Because spiritual and material aspects of living are so closely aligned in traditional Japanese thought, knowledge becomes *practice with particular forms*. Through practice, at once, knowledge is both highly formalized *and* individualized. This is true whether a novice does Tai Chi Chuan beside a master, an acolyte sits in an abbot's *dojo*, or a sage guides a student's hand across the page. What distinguishes the truly creative from the merely imitative is not the directions followed but how far and how fast one travels along well-marked pathways.

How far one travels is more a question of effort than of ability, of *doryoku* rather than *noryoku* as the Japanese put it. Obviously, there are individual differences in *noryoku* and these may be notable and important. But the keys to discovery and creativity are more dependent on effort than ability. Effort unlocks the forms set by others.

So form and effort are what count. Expression and practice within constraints are the rule. When sitting *zazen*, you become a living Buddha or so the Soto Zen sect teaches. After a thousand *osotogari* throws in countless judo *dojo*, mind and body become one in an instant flash of brilliant execution. In sitting *zazen* or in practicing judo, the pathways are the same. Do what those before you have done, only do it at least as well or perhaps just a little better. One may supersede but never displace the established forms of experience and knowledge.

The purpose of it all, from a cultural perspective, is not to succeed but to make the effort within established forms. In factories, no less than a *zendo, dojo,* or research lab, Thomas Edison's aphorism rings true: discovery is 90 percent perspiration and 10 percent inspiration. No wonder why forms are so important in Japan! The hard work has been done already because *doryoku*, or effort, is the key to creativity.

Indeed, because *doryoku* is the key, the presence of others in a work group or research team stimulates creativity. More effort, not less, is required to sit *zazen* in the presence of others. One sits in the midst of all Buddhas, past, present, and future, and that both buoys and focuses one's effort. The institutionalization of *doryoku* through group effort drives the incremental, constant process of *kaizen* on the shop floor and in firms.

Kaizen is not antithetical to *kaikaku* or radical innovation, however. Toshiba's Total Productivity (TP) movement, examined in depth in chapter 3, is closer to *kaikaku* than *kaizen*. Under the TP regime, there has been a major shift in how things are done at Toshiba. Yet once a new course is set, it is again *doryoku* and *kaizen* that develop, inspire, and carry through the discovery process. In this sense, *kaikaku* may be understood as especially intense punctuations of *kaizen* activity; both are embodied and interlaced in forms, ready to be activated by those who know how to use them. The powers of the Knowledge Works form are revealed in Figure 1.10.[36]

Hence, Knowledge Works may be a cultural artifact as much as

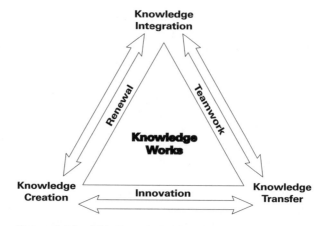

Figure 1.10   The Knowledge Works form.

an innovation architecture. For most Japanese, knowledge flows from the play of imagination with fixed, even conventional, forms. Knowledge Works are a culmination of interpreting the times, divining what is called for, and pushing the potential of past forms by learning, imitating, improvising, and discovering within existing work forms.

Knowledge Works Outline

The outline of *Knowledge Works* is as follows. In chapter 2, a general model or architecture of Knowledge Works is presented with some discussion of variations on the basic model. Chapters 3 through 6 describe different ways for moving ahead and managing the Knowledge Works model. Chapter 3 details Total Productivity Campaigning, Toshiba's way of making sure that everything works together; chapter 4 focuses on suppliers who deliver much of the value-added content of Yanagicho's products. Chapter 5 looks at Yanagicho's product development system and the world's first credit-card-sized computer; Chapter 6 highlights the many problems of taking the Knowledge Works model overseas. Finally, Chapter 7 is a summary of the model's important features with a discussion of why Knowledge Works first appeared in Japan and their international and comparative importance.

At times and in places in this book, I move beyond describing the Knowledge Works model as a grounded construct and suggest that the model is an emergent theory—a model of form and practice—for organizing manufacturing in volatile, high-technology industries. I do so because it is clear that in industries characterized by product/process complexity, rapidly changing technical and market circumstances, and time-to-market competition, it is absolutely mandatory that all of the resources and capabilities needed to compete are localized and integrated in an effective manner. While there may be a number of ways of doing this, none other than the Knowledge Works model have been articulated.

TWO

Architectures for Innovation and Renewal

Globally, Knowledge Works—multifunction, multiproduct, and multifocal manufacturing *sites*—represent best practice for the generation and application of knowledge as product. They acquire, create, accumulate, and transform knowledge into new and improved products, manufacturing and assembly processes more effectively and faster than any other model of industrial production. The quality, reliability, and speed with which these tasks are accomplished bestow obvious competitive advantages on Toshiba.

Sites are emphasized in this definition. Really valuable knowledge is "sticky" in rapidly changing, high-tech circumstances. It "sticks to" specific products and processes or, more accurately, it sticks to those responsible for designing, developing, making and marketing products and their related manufacturing processes. The more sticky the knowledge—an accumulation of countless efforts to join theory and practice in research, design, development, product manufacture, and marketing—the more it may become a potential source of competitive advantage. But in order to become so, sticky knowledge must be converted into organizational knowledge, that is, sticky knowledge must be embodied in organizational form and practice.

That happens in factories where industrial R & D and engineering practice are joined. Obviously Knowledge Works are not "factories" in the traditional sense of the word, although that is what they actually are. "Factories" are typically low-value-adding facilities where design and development activities that necessarily precede production are disassociated from sales, sales engineering, and marketing activities that follow. The main activity in "factories" is making things, pure and simple. In Knowledge Works, by contrast, an entire value-adding sequence from industrial R&D through marketing occurs on-site. The intellectual capital generated on-site is the sticky knowledge on which future profitability depends.

This full-bodied formula corresponds with the central role of knowledge creation in economic growth. As defined by Nobel prize winning economist Simon Kuznets, "an increase in the stock of useful knowledge and the extension of its application are of the essence of modern economic growth."[1] Knowledge Works fit Kuznets's model. They increase the stock and application of useful knowledge by *combining* all important industrial functions on-site, thereby recasting manufacturing as the core activity integrating industrial R&D with sales and marketing. This is the multifunction part of the model.

Collocation and integration of functions in turn enable extreme levels of differentiation, encouraging a decentralization of authority and a delegation of responsibility to the department level of organization. By doing so, by making individual departments responsible for the design, development, manufacture, and marketing of their own products—the multiproduct part of the model—Knowledge Works ensure the cost competitiveness, focus, reliability, stability, and quality of the production process, thereby confronting the traditional problems of high-volume manufacturing. Such concerns are the multifocal part of the model.

Such a model—multifunction, multiproduct, and multifocal— is appropriate for many high-tech industries, given their high information processing needs up and down the value chain, high ratio of R & D expenditures to sales, and often volatile markets. Knowledge Works are especially apt for the electronics/electrical equipment industry, which is characterized by manufacturing strategies of two sorts: one aims to improve product performance and lower manufacturing costs, and the other to diversify and refine product offerings. Thus both focus and flexibility are sought in an industry characterized by billion-dollar investments in new semiconductor "fabs" (fabrication plants) as well as by firms marketing more than 10,000 different kinds of products. Competition is all about realizing both focus and flexibility, and gaining market share by being fast to market. Toshiba's Knowledge Works model accommodates all of these requirements, and so it is today's winning, high-tech manufacturing strategy.

Administrative Heritage

Knowledge Works are not a recent, postwar invention. They appeared during the interwar period as focal factories: multifunction and occasionally multiproduct factories that bore administrative responsibility for serving regional markets at a time when national markets were not well integrated. A history of focal factories and how they were organized and managed at Toshiba is found in *The*

Japanese Enterprise System.[2] During the postwar period, focal factories were given added responsibilities for devising cost-effective responses to increasingly rich and diverse market opportunities. Additional management, engineering, and marketing resources were concentrated at the level of factories in order to facilitate fast-track feedback between markets and production. Divisional and corporate planning functions lagged as a partial result.[3] Well into the 1960s and '70s, large corporations with a decided production bias (unitary or U-Form firms) were still the rule.

What sold well at home often sold well overseas. The first Japanese cameras, motorcycles, VCRs, and automobiles to succeed in overseas markets were virtually indistinguishable from those back home. In time, however, flexibility in domestic manufacturing routines allowed for a broadening of product lines and for a differentiation of products for home and overseas. The direction of causation is important. As aggregate demand soared and manufacturing boomed along, products multiplied and became increasingly differentiated. By the 1970s, families of related products, some destined for the home market and some not, blanketed nearly every conceivable market niche and product segment. Administratively strong, technically versatile, and organizationally flexible factories provided the push behind these developments.[4]

By the 1970s and '80s, research and development activities became a key means by which firms distanced themselves from rivals; firms rushed to establish central R & D labs and to deepen product design, development, and engineering activities. Not surprisingly, the R & D emphasis sent more resources to factories. Factories were already sites of administration, production, and distribution coordination, and they were running full and steady. They already occupied prime sites in megacity markets and benefited from well-entrenched supplier networks. Factories were not bypassed by the R & D boom; instead, they were the prime beneficiaries.

During the 1970s, Japan's leading industrials pulled abreast and sometimes surpassed comparable Western firms in terms of R & D expenditures, expressed as a percent of sales or revenues. One study put R & D as a percent of sales at about the same level, 5 percent, for a sample of leading U. S. and Japanese firms in 1978.[5] Since then, however, expenditure levels have diverged dramatically. Japan's firms responded to the petroleum oil shocks of the 1970s and the devaluation of the yen since 1985 with an investment binge: investment in plant and equipment, in training and education, in suppliers and affiliates, and most notably, in R & D.

Before the recession of the late 1980s, capital investment in plant and equipment had been running as high as 27 percent of GDP (gross domestic product) and leading firms were putting 8 to 13 percent of sales receipts into R & D.[6] Even during the 1990s, Japan's industrials are still outdistancing all but the most aggressive rivals on R & D expenditures as a percent of sales, patenting activity, and technology investments. It was during this extraordinary and unprecedented era of postwar economic growth and R & D expenditure that a working distinction arose between what Toshiba calls *ryosan kojo*, or volume production factories, and *kaihatsu kojo*, or development factories. Knowledge Works are Toshiba's development factories.

The Basic Architecture

Toshiba's Knowledge Works consolidate a half-dozen distinct industrial functions—research and development, design, engineering, production, product and product department management, and marketing—in one operational structure. That is, everything from applied research, design, prototyping, tool development, testing, quality assurance, trial and scaled-up production, market evaluation and feedback is done on-site. Saying so, simply and straightforwardly, conceals the long developmental cycle that preceded this organizational architecture, the art and wisdom of it, and the strategy behind it. For organizations as well as for people, experience is a dear teacher but a good one.

A long period of trial, error, and experimentation in adapting, adjusting, and modifying different organizational forms and functions preceded the emergence of Knowledge Works. That history and administrative heritage are pivotal. Without them—without ac-

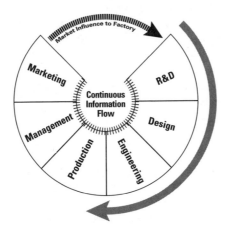

Figure 2.1 The Linked Chain or Multiple Functions model.

cumulating and refining local experience—what works best in any given circumstance cannot be identified and isolated. Because of them, a well-defined organization form emerged and evolved: it had generous yet unified boundaries and encompassed multiple functions, products, and purposes in single, integrated sites. So, while Knowledge Works, like all factories, boast a division of labor and a specialization of tasks, these are cached and collocated with complementary R & D, design, engineering, management, and marketing functions.

The 3 Ks

By corralling ample resources under one management within a unified vision and integrated leadership, Knowledge Works operate quickly, efficiently, and well, and become, in the process, imbued with an abundance of knowledge, know-how, and *kotsu* (Japanese for "knack"). Knowledge, know-how, and knack, the 3 Ks, are site-specific, that is they accumulate in specific sites as a result of particular policies and activities undertaken there.[7] The 3 Ks represent organizational knowledge, something quite different from general knowledge and something quite sticky in its generation and application. While the 3 Ks may be transferred to other sites, as happens often between Knowledge Works, affiliates, and suppliers, intangible assets like these are not transferable without effort and some loss in transit.

Hence, a basic tenet of this book is that production knowledge and practice are inseparable, and that the connection and fusion between them may be reworked, even strengthened, through repeated processes of knowledge acquisition and application. Repetition occurs in successive product development cycles and in the intense, dense, and durable transactions that bind core suppliers and factory sites together (see chapter 4 for more details). Because knowledge and practice are inextricably linked, the 3 Ks are site-specific. In other words, organizational knowledge is sticky in generation, application, and evolution.

Site-specificity

Toshiba's successful line of photocopiers are designed, developed, manufactured, and marketed at Yanagicho, *and nowhere else*, as detailed in this book. Site-specificity, in this instance, is knowing how to integrate sophisticated hardware requirements with appropriate software and sales engineering detail, and when to train, motivate, and involve employees in these activities. Only designers, developers, and operators intimately familiar with the products and production processes are able to alter specifications quickly and accurately, because such details require hands-on management and decision making.

On-site capabilities of this sort are in some ways aesthetic acts. Each new generation of products demands ever higher levels of design, development, and processing know-how, information sharing, and sense-making on the part of everyone involved. In rapidly changing technical and scientific fields such as integrated circuit design and development, manufacturing and assembly process know-how are really art-in-the-making. This is also true in other fields, such as solid state physics, three-to-five axis robotics, molecular biology, and genetic engineering—any advanced field where knowledge is as much art as science.

But art is not born in nothingness. Workmanship, intuition, feel, discipline, practice, and style, all make a difference, not once in a while but all along the value-added continuum of product design and development. Because such intangible qualities are tied to particular products and processes and to particular persons who manage product and process flow, a rich, sticky, highly textured culture characterizes factory sites with 3-K asset specificity.

What holds the mix together is Knowledge Works or the form of organizational, technical, and social resources found at particular sites. Employees are obviously the transforming force underlying and inviting the evolution of form. Their intelligence, effort, and ingenuity are what pay off. The significance of site-specific assets embodied in local, organizational knowledge cannot be ignored, and a search for these recommends the localization of functions and activities at single sites, and the training, motivating, and managing of all manner of employees there.[8]

Relation-specificity

Site-specificity and knowledge-intensivity create a need for focus, specialization, and a reasonable division of labor. These are realized by situating numerous functions for a variety of products in single sites and by nurturing networks of suppliers for developing and making products in partnership with core sites. These are relation-specific assets: good internal relations in product departments, and a network of suppliers responding to product department needs. The more advantageous the relations, the more likely it is that high performance and satisfaction will be achieved. Good relations and an accompanying division of labor allow factories to maintain product market focus while simultaneously encouraging flexibility in organizational arrangements and diversity in product/process capabilities.

Note well the differences between Knowledge Works and traditional models of manufacturing organization. This is neither a functional architecture with appended product departments nor a matrix structure with intersecting functional and product nodes. In other words, Yanagicho is not simply a strong factory organization that could make any sort of product, nor is it simply a particular division's manufacturing facility. Because the multidivisional form of corporate organization is a recent, postwar development in Japan, large, complex, and older factories such as Yanagicho typically make products for any number of divisions. In fact, Yanagicho makes products for five divisions, confounding any straightforward correspondence between functions and markets.

Given the complexity and diversity of everything that is happening, organizing principles that cut through the medley of complications and variations while preserving the site- and relation-specific nature of organizational knowledge are needed. The principles are two: largely self-contained product departments, and good lateral relations with suppliers of parts, components, and subassemblies for particular products. However, both principles may be combined in one operational rule: multifunctional capabilities—everything from design to marketing—are embedded *within* product departments and appended supplier networks, as seen in figure 2.2.[9]

The golden rule of Yanagicho's operations makes perfectly good sense. When production knowledge and practice are inseparable and where many different products are being simultaneously designed, developed, and made, product departments should have enough resources, capabilities, and authority to manage as much as

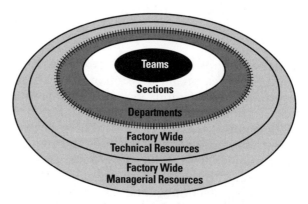

Figure 2.2 The Doughnut Ring or Multiple Products model.

possible on their own. Otherwise, the noise of a dozen different departments scrambling for favors and resources, and the difficulties of administrating them top-down from the factory manager's office, would make for an unmanageable arrangement.

Organizing for Multiple Functions and Products

Multiple products may range from numerous models of the same product, such as Toyota's full line of motor vehicles—sport coupes, two- and four-door sedans, luxury GTOs, and light utility vehicles—to products that are rather different in form, function, and features but happen nonetheless to be made at the same site, such as Yanagicho's SuperSmart cards, photocopiers, utility meters, and automatic mail-sorting equipment. Yanagicho's product diversity reflects Toshiba and the electronics/electrical equipment industry's manufacturing strategy. Hitachi's factories have even more resources concentrated in them, whereas Matsushita Electric Industrial's (MEI) are less factory focused.[10] Most of MEI's products are sourced from organizationally independent but financially dependent, original equipment manufacturers (OEMs).

An examination of 16 Toshiba factories (60 percent of the domestic total) reveals that during the 1990–91 fiscal year, half had 12 or more distinct product lines (and usually dozens of different models for each line). In some cases, an incredible 19, 20, 21, or 23 different product lines per site may be found.[11] The larger the number of product lines, the more likely a factory is serving two or more corporate divisions. Less obviously, the larger the number of product lines and corporate divisions served by single sites, the more authority they wield, the more responsibility they shoulder, and the more strategic their role in determining corporate performance.

As illustrated in Figure 1.3, the linked chain, or *multiple functions* model, is actually subsumed within the doughnut ring, or *multiple products* model. Broadband functionality is contained within department organization for the most part. That is, functional specialization is subordinated to making competitive products. So, when products are similar, really model variations on a single theme, entire factories will be geared toward making a family of products. Toyota Motor is a prime example of this. Thus, cross-functional collaboration is a question of interdepartmental coordination in Toyota's factories.

But when products are not so similar, cross-functional collaboration is realized *within* departments dedicated to the design, development, assembly, and marketing of particular products. *Intramural* rather than interdepartmental coordination is key. Electrical/

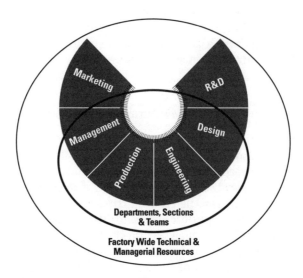

Figure 2.3   The Linked Doughnut: Multiple functions overlaid by multiple products.

electronic assembly operations are much more likely to be of this sort, as seen in Toshiba's Knowledge Works. Toshiba's product diversity and, thus, its product department robustness at the factory level of organization is mirrored in other industries, like microprocessors, nonelectric machinery, some measuring and calculating instruments, and certain telecommunications and biotechnology applications.

The frequency of multiproduct factories in such industries suggests that single-product models of manufacturing organization and operations management seriously underrepresent the complexity of the contemporary situation.[12] Multiproduct plants force a reconsideration of the effects of minimum efficient scale (MES) by shifting the calculation away from how large single-product plants should be to how many different product lines can be efficiently assembled and managed in single sites. Obviously, multiproduct plants are a more efficient use of resources and a better risk reduction strategy. More is less — less expensive and less risky.

Also when intramural resources prove inadequate, within-factory resources may be called upon in Toshiba's Knowledge Works. Additional technicians, engineers, and workers, for example, may be temporarily transferred from one factory department to another as long as more focus, specialization, and product-specific know-how are needed. However, these are temporary resources, on loan until production runs fully and steadily again. Once this happens, employees are returned to home departments or reassigned according to need.

Given boundary permeability and embodied capabilities, all product and process combinations are possible within the Knowledge Works form. New core products can be designed and developed, even as existing products receive add-ons and enhancements. Or parts and components can be redesigned, relying on factory-based and supplier-network-based capabilities. Everything from knowledge acquisition, concept investigation, prototype building and testing, pilot and full-production runs, can be accomodated within the Knowledge Works model. Not only are all of these product and accompanying process combinations possible, but also managers have the ability to choose which combinations they prefer. And quite obviously, they do so for strategic reasons.[13]

Hence, department managers have the resources and authority to manage well at their level. They are concerned primarily with the effectiveness of internal, cross-functional, information-sharing processes within their own departments. Yet department walls can be open to "outsiders," both managerial and technical, because factorywide activities, such as Total Quality Management (TQM) and Total Productivity (TP) campaigns, help ensure the permeability of department boundaries. Factory-level managers, in turn, are free to concern themselves with the overall balance of factorywide resources and activities because departments are taking care of day-to-day concerns.

Like nested yet interconnected (permeated) doughnut rings, tasks are demarcated by boundaries separating shop-floor teams from technical support sections; product departments from factory-wide research, design, and development functions; staff units from managerial units. From workers to top managers, as many as six or seven levels of organization may be found; managing so many products and functions at single sites is inherently complex. Teams and sections are controlled by department-level coordination, and departments are nourished by wide-ranging resources at the factory level of organization.

Managing Knowledge Works

Knowledge Works contain market demand and technical uncertainty by balancing the push and pull of multiple functions—R & D, design, engineering, manufacturing, management, and marketing activities—and multiple products within large, well-endowed, and integrated production sites. The art and strategy of Knowledge Works are managing boundaries and their intersection along three contingent axes—market pull, product/process complexity, and technology push—as described below.

1. *Market pull.* Production runs vary by frequency of model changes and product life-cycles. For example, in the electrical equipment industry product life-cycles may be as short as six months (radio cassettes) or as long as ten years (power generators). Given such rate differences, a factor of 20 to 1 in these two cases, a proper mix of products and an appropriate balance among them is hard to find at single manufacturing sites. (It makes a lot of difference whether products are mature or recently minted; if they are intermediate or final goods; if features or costs are more important.)

2. *Product/process complexity.* Design and systems engineering requirements vary enormously from product to product, despite similarities in appearance, form, and function. Standardizing design, parts supply, and process know-how helps to minimize such differences and to find and maintain a balance in operations. (Complexity may be measured in various ways: by the number of parts per final unit of assembly; by the degree of interchangeability in parts among models of the same product family; by the kind and amount of supplier value-added contributions to manufacturing costs; by levels and kinds of design and manufacturing automation; or by the importance of systems engineering know-how.)

3. *Technology push.* The strength of association between near-term and intermediate-term R & D and current production is an important indicator of how much technical continuity exists in upstream and downstream functions. Depending on how much products under development resemble current market offerings, linkages between functions may be extremely tight or rather loose. (Technology push depends on a lot of factors: industry maturation, length of product life-cycles, frequency of innovation, and how innovations are induced. There is far more continuity in these factors in the motor vehicle industry, for example, than in microelectronics. So, there is less technology push in the auto industry.)

Conceptually speaking, these three axes—markets, product/process complexity, and technology—generate rate differences among the many competing products, functions, and purposes of a Knowledge Works. Managers constantly monitor, mend, and modify department boundaries in an effort to juggle the competing needs of different products and development schedules. Obviously, points of balance between competing needs are difficult to find and maintain. And as soon as balance is realized, it is just as likely to be lost. Therein, the form of Knowledge Works embodies an aesthetic as well as strategic appreciation of how technology, organization, and management intersect and interact, and how they should be bound together.

Much of the balancing occurs at an unconscious level; technicians, engineers, and managers feel when things are going well. Obviously, there are countless measures and calculations, objective as well as subjective of when operations are running full, steady, and on schedule. And just as obviously, standard operating procedures as well as intuition are based on trial and error. Form is important to all of them. Form represents a stylization of practice, a reservoir of discipline and routine, and these are embodied in evolutionary refinements that are constantly appearing in practice.

Toshiba's operational answer for this high-wire balancing act is to group together products that share sufficient amounts of technical and engineering similarity, so that one product's gain or slip, triumph or snafu does not bring down the act and actors. Knowledge Works spread risk across a family of products, therefore, quite unlike a single-product factory's solution to the problem of dynamic balance. Products should be sufficiently similar, either in how they are made or in how they function, to be grouped together effectively. It also helps if marketing and distribution can be carried out in similar ways.

There are two tricks and one caveat to this solution. The tricks should be familiar by now: where functional and product boundaries are drawn, and how permeable they are. The caveat is the past, which is less familiar as a solution to problems of the present but no less important. History cannot be ignored when drawing boundaries or managing the deployment of resources because today's policies are no more than reflections on past experience. The combination of multiple functions (linked chain model) and multiple products (doughnut ring model) yields factories that are socially and technically complex, yet functionally flexible, versatile, and integrated.

This synthesis, so simple in concept, so widespread in Japan, is unrivalled worldwide because it reflects a country-specific pattern of organizational evolution: late development, the focal factory model, extraordinary concentration of wealth and consumption in the urban corridor joining Tokyo and Osaka, and unprecedented economic growth and labor-management accord during the postwar era. Knowledge Works, in other words, evolved as sensible solutions for coping with rapid technical, market, and environmental change. Yet, insofar as these conditions are widespread, the model is generally applicable in places like New York, Los Angeles, London, Paris, Tokyo, Seoul, and Shanghai.

Tokyo Effects

In every great age and civilization, cities are the epicenter of change.[14] Rome was home to the early Holy Roman Empire; Peking has been the hub of China since the T'ang and Sung dynasties; London was the nerve center of England's vast colonial and commercial domain and Paris of France's, and New York is the financial and commercial capital of 20th-century America. Tokyo and the Osaka-Kyoto-Nagoya corridor power Japan toward the 21st century, like the capitals of old.

Knowledge Works are growing in numbers and significance in megacity Japan, where transportation, communication, and the information infrastructure are concentrated. Tokyo-Yokohama is the largest conurbation in the world, with some 28 million people in 1991—as large as the population of many nations and far wealthier. The Osaka-Kyoto-Kobe conurbation, with 14.8 million, is the sixth largest urban area in the world.[15] The scale, scope, and sophistication of these cities and markets dictate that production organizations respond with agility to their pent-up and variable demand.

"Tokyo Effects" refers to the impact of megacity markets like Tokyo on manufacturing organization, even though the impact is obviously not confined to Tokyo alone. The point is that extreme environments call for extreme responses, and in large part Knowledge Works are an organizational response to "Tokyo Effects." The accumulated impact of Tokyo Effects on Japan's major domestic firms for the past three or four decades means that they are now ready to respond to the incessant pace of global change. Tokyo Effects have already thrust Toshiba and other major firms into time-based competition in product design and process development at home. Getting to markets early with the right products at the right prices reaps most of the profits in fast-to-market competition. By the time fast-follower firms get there, consumer preferences have changed and most of the profits have disappeared.[16]

No doubt California's Silicon Valley and Boston's Route 128 are important and demanding markets in their own right. Yet they perch on opposite ends of a great continent, separated by 3,000 miles, in competition with a host of lesser but nevertheless important cities, the Seattles, Denvers, Cincinattis, and Charlottes of the world. Little more than 300 miles separate Tokyo and Osaka, and with all due respect to Osaka's "we try harder" campaign, Tokyo outdistances Osaka in terms of the numbers and significance of government offices, industrial research labs, Fortune 500 companies, university students, newspapers, radio and television stations, fashion designers, financial consultants, foreign firms, and just about any other measure of fame and fortune that one might wish to apply.

While Tokyo outshines Osaka, the Osaka-Kyoto-Kobe conurbation is nonetheless huge. The three largest metropolitan regions, Tokyo, Osaka, and Nagoya, grew by more than 50 percent between 1960 and 1980. Including these cities as anchor points in a vast megacity, high-tech core, 50 percent of Japan's 130 million people now live, work, play, and consume in a single, vast, unified market.[17] In a single day alone, about 20 million move in and out of Tokyo.

Because Toshiba's roots and most of its facilities are found in the southwestern districts of greater Tokyo, including Kawasaki and Yokohama, Toshiba is unusual in the degree to which it is driven by Tokyo Effects and in the geographic proximity of its many research, design, production, and managerial units. Kawasaki City, home of the Yanagicho Works, is criss-crossed with Toshiba sites. Toshiba has 8 of its 15 corporate and divisional research facilities in and around Kawasaki City. These include the prestigious Central Research and Development Center, the Manufacturing Engineering Laboratory, the Very-Large-Scale-Integration (VLSI) Research Center, and the Systems Software Engineering Center.

The accompanying map in Figure 2.4 details the concentration of Toshiba's R & D, design, and production in Kawasaki City, Yokohama, and the Shibaura district of Tokyo. Toshiba's spatial concentration replicates a basic strength of Yanagicho: an integration and localization of functions and capabilities. Kawasaki City and southwestern Tokyo are Toshiba's Silicon Valley, a local area network of concentrated high-value-added activity. Less trend-conscious and time-sensitive products, like refrigerators, air conditioners, and the like, are manufactured outside of the zone of Tokyo effects, either in the countryside or increasingly off-shore in places like China and Malaysia.

Positioning integrated design, development, production, and marketing systems, like Knowledge Works, within demanding markets such as greater Tokyo, Kawasaki, and Yokohama, create an organizational architecture that is off-the-grid in terms of traditional manufacturing models.[18] Traditionally, production stressed command and control functions: managers dictated what would be made, and when and how; products were pushed onto the market rather than being pulled by the market. But command and control economies, even when they work well, produce extreme Taylorism, lots of bureaucracy, and a spotty supply of mediocre products.

Knowledge Works and Tokyo Effects make sense together. Tokyo is a zone of extreme economic behavior, an example of what is sometimes called "neighborhood effects": consumption is telescoped, transactions are accelerated, already specialized markets are split into finer, ever more demanding segments. Only the best, the

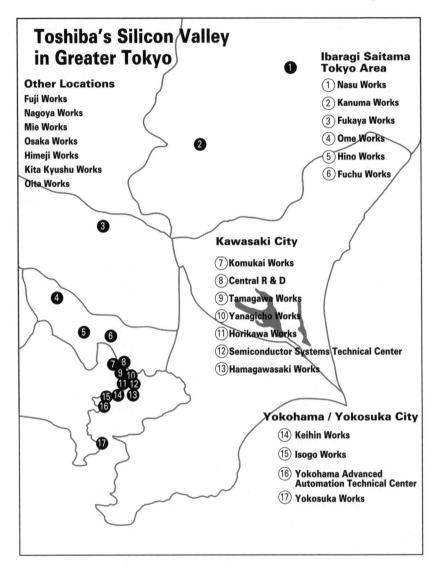

Figure 2.4 Toshiba's Silicon Valley in Greater Tokyo.

latest, the zaniest, and the most up-to-date products sell well. Casio, for example, changes its radio cassette product lineup every three or four months; Toshiba photocopiers have a six-month product life-cycle; Toyota, Nissan, and Honda swamp GM, Ford, and Chrysler with full-model changes every four years compared with their five to seven. Fall behind Tokyo's consumption curve, fall behind Tokyo's emphasis on product quality and diversity, and you are unlikely to catch up.

In other words, Tokyo and its megacity sisters are the world's most affluent, concentrated, and trend-conscious markets. In the years 1957–72, years that the Japanese call "the age of high-speed economic growth," national wealth grew 2.8 times, real personal consumption 2.2 times, while the cost of food stuffs as a percent of household expenditures fell from 45 to 32 percent.[19] By the early 1970s in Japan, 100 percent of households enjoyed black and white televisions, nearly 90 percent had refrigerators and electric fans. By the early '80s a second wave of consumption put color televisions, vacuum cleaners, answering machine telephones, and automobiles into nearly everyone's hands.[20]

A third wave in the late '80s gave Japanese consumers multi-colored Walkman, 32-bit laptop computers, camcorders, cordless phones, fuzzy-chip rice cookers, Miata roadsters, and toilet seats with built-in heaters and fans. Throughout the high-growth years and beyond, into the years when Japan outdistanced but did not bury the competition, living standards rose much faster than urban population totals. The waves of consumption were nowhere more exaggerated than in Tokyo, although Osaka, Kyoto, Kobe, and Nagoya were not far behind. Toshiba's success and Yanagicho's organizational form were and are location-dependent to a considerable degree.

Innovation Architectures: Champion Line Model

Because certain products and processes are more important than others in Knowledge Works, balance of functions, products, and goals is not a mid-point, a mean, an average of everything. Some products, processes, and purposes are simply more important than others. Importance may be measured by value-added contribution to manufacturing output, the amount of dedicated floor space, or the number of employees attached to product departments. But in Knowledge Works, intangible assets like design and engineering know-how are more critical than tangible ones like floor space and employees, so history is important for understanding how certain products/processes have been critical in the evolution of a plant's engineering, assembly, and managerial knowledge.[21]

For example, during the 1960s and 1970s, when televisions, refrigerators, washing machines, calculators, and pocket radios were sold in huge volumes, scale-oriented manufacturing experience endowed factories with product and process-specific resources and capabilities. In order to run full and steady with very little slack, design and engineering resources were dedicated to making particular products in huge volumes. Such products, say, calculators or tran-

sistor radios, assumed unusual importance in the evolution of a factory's knowledge endowment. In other words, organizational learning is not an abstract activity but a concrete one.[22]

Yanagicho's endowment of knowledge-based resources and capabilities, that is, its technology management setup, systems engineering know-how, and product/process mix, are rooted in particular manufacturing experiences. PPCs (plain paper copiers) have been
unusually important in the evolution of Yanagicho's organizational
knowledge. The weight of the PPC line was approximately 50 percent of total production value during the 1980s, and this decided dependency on photocopiers underscores the practical difficulties of
balancing organizational resources across a spectrum of products and
functional capabilities.

Take PPC and laser beam printers (LBPs), for example. When
175,000 PPCs are being assembled annually at Yanagicho, what it
takes to make top-flight PPCs is not really what it takes to produce
30,000–35,000 high-quality LBPs, in spite of obvious technical similarities in the two products. So, while it is possible to link the design, development, and manufacture of PPC and LBP, it is not at all
desirable to do so in fact. In terms of volume alone, five to six times
as many PPCs are assembled as compared with LBPs. While the correspondence between manufacturing volume and the amount of
dedicated resources needed to produce in volume is not one-to-one,
sustaining five to six times higher volume of PPC assembly obviously means that larger amounts of Yanagicho's multifunctional capabilities—everything from applied research, design, prototyping,
and process engineering to operations and production management
know-how—will be dedicated to PPC manufacture.

Also, key components and product strategies diverge dramatically for the PPC and LBP. While selenium drums are a key component for high-performance PPCs, laser diodes are for LBPs, and
printed circuit board layout, design, and functions differ for the
two products. PPC product life-cycles are about six months long in
Japan, that is, within half a year rival firms will bring out similar
products that compete favorably on price and features. LBP product life-cycles are two or three times longer, given a less mature
market for laser beam office equipment. So, other than production
volume differences, a number of key design, development, and
planning differences distinguish PPC and LBP manufacture, and
this is true for two products that are more technically similar than
most.

Although the PPC and LBP are not so alike in development,
planning, and product/process requirements, the fact that PPCs are
made at Yanagicho guarantees a scale of activities in applied research, design, development, engineering, tool and diemaking that

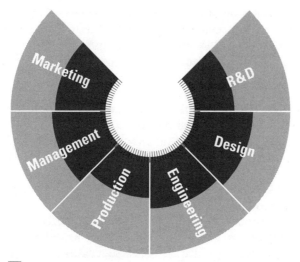

■ The amount of available resources that are dedicated, hypothetically,
to the development and production of a champion-line model.

Figure 2.5   Champion Line model.

enriches and nourishes the development of related products, including LBPs. Scale and volume, pure and simple, count for a lot because a certain level of activity must be realized before knowledge,
know-how, and knack accumulate and are transferred for spillover
into other product areas.

In other words, PPC design, development, and assembly became
the knowledge base for designing and making LBPs, ATMs, automatic mail sorting equipment, and a host of related products at
Yanagicho. For knowledge spillover and transfer to occur, however,
a threshold of scale-oriented, multifunction activity had to be realized in one product area before the experience and accumulated know-
how could benefit other product areas. Products that have gained
that level of investment and proven to be a reservoir of knowledge
and know-how spillover to other products are called Champion Line
products at Yanagicho. Champion Line products drive the organizational and technical evolution of Knowledge Works. They lay a
threshold of organizational knowledge on which related products
can thrive, one that ultimately justifies the high risk and cost of creating multifunctional capabilities in particular product departments.

Figure 2.5 illustrates the disproportionate resources dedicated to
Champion Line products. All along the multifunction value chain,
Champion Line products garner a disproportionate amount of available resources and, managers hope they generate a disproportionate
share of factory-based capabilities and know-how spillover.

Toshiba photocopiers have to contend with Canon, Mita, Minolta, Sharp, and Fuji Xerox PPCs. Toshiba has to be as good or nearly as good as they are in PPC design, development, and manufacture, even though all of these companies depend more on photocopier sales than Toshiba does. This competitive reality translates into a strategy of creating Knowledge Works-like centrality and capabilities without which Toshiba would surely have to abandon the PPC market. But even PPCs are not islands onto themselves. Once sufficient R & D, design, engineering, and managerial resources are concentrated on making and selling PPCs, a notable wealth of resources and talent have been assembled. Without a constant demand for these, an unlikely circumstance, slack resources can be deployed elsewhere as demand and opportunities arise.

Photocopier manufacture was the scale-oriented experience that allowed Yanagicho to gather and enhance its Knowledge Works-like centrality, resources, and capabilities. Successful, high-volume PPC manufacture paid the bills for the considerable investment in plant, equipment, personnel, and know-how that were needed to be cost and quality competitive in a dozen different product areas. Indeed two-thirds of the 13 different product lines made or assembled at Yanagicho in 1995 are historically and technically linked to its PPC Champion Line: cash handling machines; ATMs; bank note counters, sorters, and counterfeit detectors; optical-electric data storage devices; railroad and motorway ticket validating equipment; and laser beam printers. All of these products can be traced lineally to the late 1960s' decision to locate PPC design, development, and engineering capabilities at Yanagicho. All of these are likewise joined at the hip to the precision machining, optical-electric, electromagnetic, paper handling, image processing, software, and engineering technologies that they share in common with the PPC.

Photocopiers are still Yanagicho's Champion Line, even though they are outclassed by utility meters and refrigerator/freezer compressors in volume. Yanagicho makes about 175,000 PPCs a year compared to twice as many meters and five times as many compressors. But meters and compressors demand little of Yanagicho's research engineers, designers, and skilled technicians. Compared with the PPC, their product life-cycles are longer and their systems engineering requirements less complex. In short, PPCs have considerably more multifunctional requirements than either meters or compressors, and more important, they have spawned a school of related products and processes, making Yanagicho into a multiproduct manufacturing facility and elevating the PPC line to Champion Line status.

Balanced Line Model

Unlike the Champion Line model, where a single product dominates a factory's resource endowment, the Balanced Line model aims to spread a factory's resources across an entire landscape of products. By allocating resources to each product department on a roughly equal basis, risks and rewards are balanced across a portfolio of product offerings. This strategy gives management a way to cope with market uncertainty and changing technical circumstances.

At first glance, the Balanced Line model appears similar to the "focused factory" concept of product shops, sometimes known as the plant-within-plants (PWP) model. But it is important to distinguish Knowledge Works from the PWP concept. In the PWP model, resources are modularized, that is, decentralized around specific product shops. While some integration of general activities, such as purchasing, transport, and information processing, may be realized in this model, applied research, design, development engineering, quality assurance, testing, training, and prototype engineering, are never collocated and interrelated on-site, as in Knowledge Works. Hence, the PWP concept neither incorporates the boundary spanning mechanisms, the broadband functionality, and the factorywide emphasis on intelligence and innovation of the Knowledge Works model nor does it attempt to capitalize on these for a range of related products.[23]

The Balanced Line model (as opposed to the PWP model) joins industrial R & D with design, engineering, manufacturing, and management, but this multifunction strategy does *not* come at the expense of any product department. Unlike the Champion Line model, resources and capabilities are not disproportionately skewed in the direction of favored products. So, Yanagicho can make as few as 20 automatic mail sorting machines each and every month, not to mention 100 optical electric data storage systems, 200 automatic railroad fare systems, and 300 ATMs and other cash handling machines.

Given that a balancing of multiple functions and products is often precarious, Knowledge Works cannot function effectively without boundaries—well defined if permeable boundaries to keep product/process requirements and capabilities intact.[24] The larger the number of product lines and the more robust factorywide capabilities, the more critical internal boundaries become. In the Balanced Line model, product departments harbor enough design, development, and managerial capabilities to get by on their own. Stability and symmetry are key. Functional requirements are more or less in balance with the allocation of available resources, as illustrated in Figure 2.6.

Yet, functional requirements and resource deployment patterns

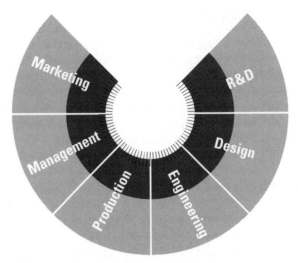

The amount of available resources that any one product requires for development, production and marketing

Figure 2.6   Balanced Line model.

are not static. For example, the PPC Department may want further testing of an engineering sample of selenium drums; the Optical Electrical Disk Drive Department may need more research on media substrates for its disk drives; yet another department seeks training in application software for sales engineers. All along the value chain, product and functional requirements and resources vary, and those in charge of getting and deploying them will necessarily compete for a larger share. Because needs and resources are not always in sync, especially in a multifunction, multiproduct manufacturing environment, there are jealousies and rivalries inherent in the Balanced Line model that are not so apparent in the Champion Line model.

Hence, the Champion Line model is more stable over time. Resources are more obviously weighed in favor of particular products and processes; such priorities free department and factory managers from the time-consuming and contentious tasks of trying to balance capabilities, prioritize demands, and negotiate resources on a more or less equal basis. And the optimal configuration of factory resources really depends on whether or not key components are designed in-house or outsourced, and that contingency only complicates the difficulties of balancing needs and resources. (Chapter 4 discusses in detail supplier relations, product management, and manufacturing strategy.)

Because of these difficulties, at least 18 different formal planning meetings occur routinely within the Information and Communica-

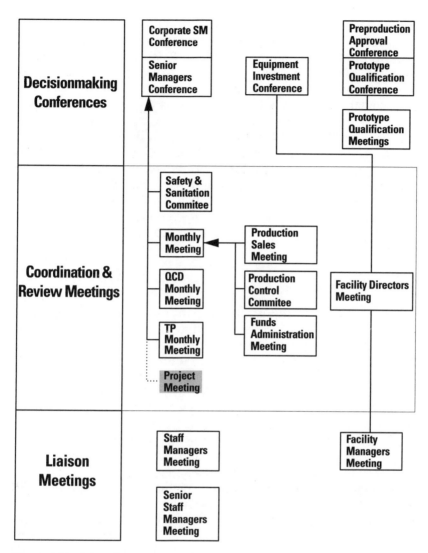

Figure 2.7   Administrative coordination at Yanagicho.

tions Group at Yanagicho (see Figure 2.7), and this ample amount of planning is entirely devoted to negotiating and coordinating the allocation of factory resources. Maintaining balance is obviously not so easy. Perhaps this helps explain the 1,270 telephone and fax addresses at Yanagicho, about one address for every three employees.[25] However, in all fairness, it should be understood that Yanagicho's organizational architecture is a combination of the Champion Line and Balanced Line models. Obviously, PPCs and closely related, high-volume products, like LBPs, garner most of the resources, leaving the low-volume lines, like automatic mail sorting equipment and ATMs, to divide remaining resources more or less equally.

Both models—Champion Line and Balanced Line—are inherently unstable in the long run, given Tokyo Effects and rapidly changing market and technical circumstances. Indeed, there may be a regular oscillation between the two as supply and demand for particular products fluctuate according to market pull, product/process complexity, and technology push as detailed earlier. Yet it is always easier to have a fulcrum for planning, coordinating, and managing, and the Champion Line model, with its best-selling products, core technologies, and disproportionate importance, dominates in this respect. Yet even this bias does not diminish a need for constant reorganization of resources and reconfiguration of capabilities, as the next chapter on Total Productivity campaigning makes clear.

Factories and Core Competencies

In today's high-tech industrial competition, factories—not corporate planning offices—determine what an industrial firm is good at. Factories define the boundaries of capabilities and competence building because today's electronic products do not embody a single discrete and homogeneous body of knowledge. Virtuosity in PPC design and production, for example, turns on a number of heterogeneous but interdependent bodies of knowledge, such as electrical engineering, chemical engineering, laser optics, and physics. For such reasons, learning and capabilities are not easily transferable or imitable across Knowledge Works, and corporate planning offices are not fully conversant with the site- and relation-specific nature of organizational knowledge. Chapter 7 details some of the difficulties that are likely to ensue from this deficiency.

It is the systemic, site- and relation-specific character of resources and capabilities, their scale, scope, centrality, proximity, and intangibility—their stickiness—that imbue Yanagicho with its Knowledge Works-like features. Toshiba's manufacturing strategy tries to take advantage of such embedded features by researching, designing, and producing families of related, rapidly replenished products. This full-line, quick-turn strategy hinges on the boundary spanning and systems integrating capabilities of Knowledge Works. Such capabilities—seen in the technical artistry, cost and performance leadership, design quality, and manufacturing reliability of Toshiba products—are mutually reinforcing. They are bundled together.

Bundling assumes mechanisms to consolidate and mobilize resources across functions while, at the same time, fixing reasonable limits to those processes. Process limits are defined in part by how well products sell. Factories also set process limits by deciding who and what to bind together in work teams and development projects. The Knowledge Works' strategy aims to consolidate six distinct manufacturing functions for numerous products in one dynamic organization. Admirable economies of scope, speed, and systems engineering arise and something approximating real-time adjustment of design and development with manufacturing and marketing emerges as a result. An effective integration of these economies occurs only when ambitious and experienced employees are motivated and well managed, and hence it is the embodiment of all this capability, virtuosity, and ambition that is Knowledge Works' greatest accomplishment.

Some Good and Bad Examples

Mitsubishi Electric's Silicon Cycle

Dynamic random access memories (DRAMs) provide a perfect case in point. It takes two years and anywhere from $0.75–1.25 billion dollars in capital investment to build a state-of-the-art DRAM production facility today. However, a product generation of DRAMS lasts a scant three or four years. Yet, if enough money is invested, if construction is on-time, and if no glitches interrupt production, investment in a DRAM factory may be recuperated in one year, enough money made in a second year to bankroll a new production site, leaving a third year of profit-making before the silicon cycle starts anew.

In the case of Mitsubishi Electric's Saijo Plant on Shikoku Island, the Saijo Number 1 Plant was built at a cost of $330 million in 1985; Saijo 2 was completed for $425 million in 1989; and a new third-generation plant was erected at Kochi in 1992 at an estimated cost of well in excess of $600 million. The silver lining to this extraordinary string of capital expenditures is MEI's progress at automating mixed-load, mixed-model IC chip production. Saijo 1 fabricated 5 or 6 different kinds of semiconductors, Saijo 2 15, and Kochi 50![26]

Hence, Mitsubishi Electric's manufacturing process allows the company to spread the risk of producing semiconductors across an increasingly larger number and variety of semiconductors. In other words, financial exposure is absorbed by and through manufacturing flexibility. Cross-functional integration of an entire range of product development skills and resources is a key feature of Mitsubishi Electric's success at Kochi. By integrating design, engineering, manufacturing, quality, and production management functions in one site, the risk of failed implementation due to ineffective technology transfer and information exchange drops dramatically.

Mitsubishi Electric's semiconductor strategy well illustrates a foreshortening of strategic choice under the impact of tumultuous markets, complex and perishable technical resources, extreme financial exposure, and incessant competition. At the blueprint level, there is little apparent difference among leading chipmakers in Japan. They all produce state-of-the-art engineering drawings and the most recondite points of hardware and software design are quickly diffused among elite computer engineers. But at the level of implementation or process engineering, few sites yield perfect chips at anywhere near a one-to-one production rate.

As a result, a 5 percent increase in yield results in much more than a 5 percent increase in profits. Although R & D labs can develop new semiconductor materials, chip designs, and microprocessor architectures, they cannot improve process yields. This is the profitable and critically important work of production sites. Such capabilities are based on organizational knowledge and they imply organizational limits; these are site- and relation-specific.[27]

Too Much Depth/Not Enough Breadth at GM and United Technologies

In the 1980s, General Motors failed to build in product flexibility at the plant level with disastrous results. Although the Chevrolet-Pontiac-Canada (CPC) group within GM was reorganized in 1984, functional, chimneylike structures were still the result. Line managers reported to plant executives in one area, while engineers in the same area reported to engineering managers. Assembly plants were dedicated to single products, and sometimes to single models of the same product.

CPC's Doraville, Georgia, plant, for example, built only Oldsmobile Cutlass Supremes. When demand for Cutlass Supremes fell in the late 1980s, the plant was already running at half-capacity.[28] Without massive investment and a complete overhaul of plant structure, other chassis and drivetrain models could not be built at Doraville. And the same dismal choice pitted GM's Ypsilanti Plant in Michigan against its Arlington, Texas, facility. Both plants were able to produce only one auto line, and when demand for these particular models fell, entire plants had to be closed down.

Although less dramatic than GM's massive retrenchment of 74,000 employees, United Technologies' loss of 13,900 jobs to restructuring was motivated by similar rigidities in factory structure. UTC's Carrier Corporation subsidiary closed its City of Industry, California, factory, which was dedicated to the manufacture of small air conditioners. While the factory's 400 employees represented a small percentage of UTC's total job losses, the plant's concentration on small-capacity air conditioners meant that it was unable to respond flexibly by altering the composition and mix of its products.[29] Rigidities in factory organization and corporate strategy cost UTC jobs, profits, and its market lifeline at the City of Industry plant.

Comparing Factory-Based R & D in Japan

Organizational differences such as these powerfully influence manufacturing strategy, as laptop computers demonstrate. Toshiba's Ome Works, on the western fringes of Tokyo, assembles more than a million computers and computer knock-down kits (KDK) annually, including the half-million laptops destined for the U.S. market.[30] In order to achieve volumes of this magnitude and stay on the cutting edge of new product design and development, the Ome Works is closer to a research, design, development, and assembly center than a traditional factory. Like Yanagicho, Ome is a Knowledge Works.

Two-thirds of Ome's 4,000 employees are technicians, designers, and engineers, while another 100 are deployed as researchers from the Information and Automation Systems Divisional Engineering Lab. In addition, about 2 billion yen is invested annually in upgrading existing facilities or in building new capabilities. Continuous investment of this sort with human resources of this caliber transforms the factory into a strategic resource for the design, development, and manufacture of personal computers. Not surprisingly, it also makes Ome into one of Toshiba's leading factories for CIM (computer integrated manufacturing) technologies.

But mobilizing resources does not necessarily result in a lot of fixed automation. The market for personal and laptop computers changes too rapidly for that. Instead, Ome Works, like Yanagicho, is more of a design and development platform and an assembly factory than a manufacturing facility. As much as possible, unit production is modularized and outsourced, and most of the assembly of parts, components, and subassemblies is done by suppliers and OEM (original equipment manufacturers) makers. The Ome end of the line may have as few as 10 persons performing 10 process steps for final PC assembly. With short-line production setups, a laptop PC at each work station gives immediate feedback on the pacing of production: whether the line is running ahead of or behind schedule; what process steps have been completed; and what is left to be done. Given this arrangement, changing the pace and content of work is simply a matter of rearranging short-line setups and changing the tracking software.

The high levels of process efficiency are related to product design strategy. In order for Ome to be mostly a design, development, and assembly site, suppliers and OEM makers have to be convinced of the benefits of producing for Toshiba. Given that Toshiba's laptop architecture is open, employing Intel chips and IBM-compatible operating systems, there is a strong argument for both hardware and software producers to climb on the Toshiba bandwagon. With an open architecture and with 1 million units being produced annually, there is a lot to gain by supplying Toshiba. Toshiba's in-house capabilities make it a desirable development partner with Intel, for example. Toshiba was the first company to bring to market a Pentium-chip-based laptop. Announced at the Comdex convention in Las Vegas in 1994, Toshiba was first to market because it had the in-house capability to design and develop the chip set that accompanies a Pentium processor.

In contrast, NEC has pushed a closed architecture of its own design. No wonder NEC had to invest a lot more in support of third-party software developers to write for its machines, and no wonder that NEC has failed to establish a beachhead for its products outside of Asia. The argument applies to suppliers of NEC parts and components as well. Suppliers are reluctant to make significant transaction-specific investments when payoffs are problematic. When NEC dominated markets, like the PC hardware market in the 1980s, suppliers were willing to support what seemed like a local monopoly. But when Toshiba's Dynabook and its imitators are smashing NEC's monopoly, and when MacIntosh, Compaq, and Hewlett Packard are making significant inroads into academic and corporate marketplaces, writing proprietary software for NEC is much less appealing.

As of 1991, comparing the three general electrical equipment/electronics rivals (Toshiba, Hitachi, and Mitsubishi Electric), Toshiba had 13,000 R & D researchers to Hitachi's 12,100 and Matsushita Electric Industrial's 7,000. This puts about 19 percent of Toshiba's employees in the R & D game as compared with 15 percent for Hitachi and Mitsubishi Electric, reflecting Toshiba's emphasis on getting more employees involved in R & D activities lower in the organizational hierarchy. And because more lower-level employees are involved in research and development, R & D costs for Toshiba are less than what might be expected. Toshiba lags behind Hitachi and breaks even with Mitsubishi when it comes to R & D investment. Hitachi spends some 9 percent of sales on R & D while Toshiba and Mitsubishi spend about 7 percent.[31] So Toshiba has more people doing R & D for less than either Hitachi or Mitsubishi Electric.

Organizationally speaking, and this is really the point, Toshiba arranges its R & D activities rather differently from the other two. While all three boast a single, central R & D laboratory, at the divisional level of organization Mitsubishi has 11 sectoral research facilities, Hitachi 9, and Toshiba only 4. In other words, Mitsubishi Electric and Hitachi have chosen to concentrate their R & D activities at relatively higher levels in the organizational hierarchy than Toshiba. This is consistent with orthodox thinking on the purpose and position of R & D in the corporation. Orthodoxy holds that basic and applied research are highly specialized activities that necessarily occur upstream from actual process and production engineering. Accordingly, basic and applied research are best isolated from the daily hustle and bustle of operations management, product planning, and marketing. Most research labs, even applied ones, are far removed from the mundane world of everyday manufacturing and management.

Toshiba stands this orthodoxy on its head. At the factory level of organization, Toshiba has ten so-called engineering or Works Labs, while Hitachi has two and Mitsubishi has one. In other words, Toshiba concentrates its R & D resources much lower in the organizational hierarchy than Mitsubishi Electric and Hitachi do. The latter two choose to focus R & D resources at the top and mid-range of the organizational hierarchy, following the conventional pattern of separating out specialized research, design, and development activities from day-to-day management of operations.

By contrast, Toshiba concentrates 75 percent of its R & D personnel at the factory level of organization and another 15 percent at the divisional or sectoral level, thereby consolidating 90 percent of its R & D workers below the level of central R & D. This unorthodox organizational design integrates top-down applied research and product design activities with factory-based, bottom-up approaches to product/process engineering. As long as Toshiba is able to capture, consolidate, and integrate most of the knowledge and know-how that flow from this design solution, it appears to shorten time-to-market speed and capitalize on learning opportunities up and down the value chain of applied research, design, development, and manufacture of new products.

In the fall of 1990, for example, the official roster of Yanagicho employees totalled 3,263 employees, 2,758 of whom were regular factory employees and another 505 of whom were so-called resident factory employees. The distinction is important. Regular employees are assigned to and paid by Yanagicho while resident employees are assigned to various Works Laboratories and engineering research units (but paid by the divisions to which they are attached). Through this arrangement, medium-term R & D activities of divisions can be conducted at manufacturing sites, facilitating the flow of know-how and personnel between R & D and manufacturing functions.

In fact, the loading of Yanagicho's roster with knowledge workers is even more pronounced than the official headcount of 505 resident factory employees suggests. In August 1988, for example, there were actually 1,805 Toshiba employees in Yanagicho's Power Tower. Of these, the largest number, 1,114, were regular employees but the remainder, 691, were resident-unit employees concentrated in three areas: telecommunications and information systems research; software systems and software engineering; systems engineering and sales engineering.[32]

In addition, resident engineers from affiliates and suppliers are always visiting the factory; further, in the summer of 1988 there were 126 part-time employees and 286 design engineers from Toshiba Intelligent Technologies assigned to Yanagicho, not to mention the dozens of engineers from Toshiba Chemical, Toshiba Silicon, Toshiba Battery, Marucon, and elsewhere working on the smart card development project, for example. An assortment of long-term visitors, like myself, from all kinds of places with all kinds of purposes are also present. At any time, around 4,000 highly skilled and motivated researchers, engineers, technicians, managers, and workers can be found at Yanagicho. In other words, the official roster of 2,758 regular employees only represents about two-thirds of the actual number of persons working there, and almost all of the additional "irregulars" are university-trained scientists, engineers, and managers.[33]

In the 1990s, Toshiba's organizational design is the winning strategy. Integration of top-down and bottom-up information processing, knowledge, discovery, and know-how generation is the key to winning in time-based competition. This requires a simultaneous strengthening of the industrial R & D lifeline, extending downward from central and divisional R & D as well as upward from factory-based research, design, development, and production activities. Distinctions between basic, applied, and development research need to be maintained, even as the time and effort to traverse from one research realm to another needs to be shortened. Toshiba's 4 corporate labs with 10 percent of its R & D workers, 10 engineering labs with 15 percent of R & D employees, and 27 factories with the remaining 75 percent highlights its balanced approach to R & D organization. Recall that engineering, or Works Lab, and factory personnel are collocated, physically and organizationally, in Knowledge Works, though on paper (on the organization chart) a distinction is made between regular and resident employees.

Arguably, Toshiba has the most balanced approach among all the general electrical equipment/electronics makers. Toshiba's posture represents a higher level of adaptation at a lower level of organization. Too high a loci of R & D and factories and development projects benefit only indirectly from centralized research. Too low a loci,

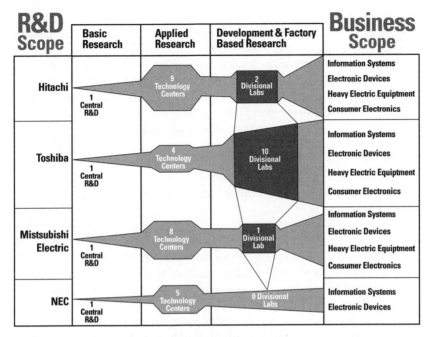

Figure 2.8 R&D organization and scope of business:
The electronics/electrical equipment industry.

on the other hand, and short-term business goals are likely to be emphasized at the expense of innovation and discovery. Balance is crucial. Knowledge Works are designed and managed with notions of dynamic balance in mind, although it is clear that balance is often hard to find.

Figure 2.8 illustrates some of the differences between Toshiba's R & D design strategy compared with Hitachi, Mitsubishi Electric, and NEC. The comparisons focus on scope, or the breadth, of R & D activities in light of the breadth of business activities. Business scope is calibrated by four major product areas, namely communications and information technology, electronic devices, heavy electrical apparatus, and household electrical goods.[34]

The M-Form Factory versus the M-Form Firm

Knowledge Works emerged in Japan during the postwar era of double-digit economic growth, 1959–1973, and developed thereafter under the twin stimuli of export promotion and domestic economic expansion. Knowledge Works embody a realization that manufacturing organizations have to change at least as rapidly as social, economic, and political circumstances do, yet changes are always limited by what is available: the mix of specific technologies, standards, products, and processes that constitute a factory's accumulated store of sticky, organizational knowledge.

If an economy grows at 10 percent a year, if 1 megabyte chips can make 256-K chips obsolete in a single swift stroke, if voluntary export restraints make 20 percent of a factory's output redundant, how can organizations respond with agility and accuracy to such circumstances? In this topsy-turvy world, organizational knowledge allows industrial firms to deliver a full range of products quickly, economically, and well by choosing what is appropriate from what is locally available in factories. Like mythical Prometheus, Knowledge Works harbor the ambition, resources, and energy to labor against capricious fortune.[35]

Knowledge Works do so by customizing production, emphasizing product variety and volume, shortening time to market, and by focusing on learning as the ultimate organizational ambition. Such interlocking workplace features, emphasizing speed, flexibility, and the inherent intelligence of modern factory workers, transform the nature of work and hence the character of industrial competition. The transformation of industrial production and competition has proceeded further in Japan than anywhere else in the world, leading to a division of factories into two sorts: those stressing innovation, flexibility, and fast-to-market manufacturing capabilities, and those favoring longer, more stable and standardized manufacturing runs where low costs are the rule.

Traditionally, low-cost, high-volume factories were located in Japan's countryside. Today, they are increasingly located abroad in places like Thailand, China, and Mexico. The other sort of factories, called Knowledge Works herein, have always been located in the epicenter of change, the megacity corridors linking central Japan. Yet, today, with the yen's high value, global competition, and the rapid pace of technical progress, Knowledge Works' resources and capabilities that are not so site- and relation-specific are being moved offshore. However, this is more easily said than done, as argued in chapter 6.

Knowledge Works couple upstream and downstream functions for multiple products and engineering processes, manage their co-evolution, unite them in a fertile organizational culture, and attempt to move it all ahead. Creating, reworking, and applying intellectual capital is the Knowledge Works model. In the West, by contrast, such tasks are typically the responsibility of product division managers in multidivisional or M-Form firms. Multidivisional firms are related-product firms that manage a variety of assets (skills, technology, and knowledge) that are subject to excess capacity. In order to make fuller use of these assets, M-Form firms separate out operational from strategic decisions, giving responsibility for the former to product divisions and for the latter to the headquarters. In this arrangement, basic research is typically a corporate responsibility while applied research and advanced engineering are divisional ones. Factories, several levels down the organizational hierarchy, are usually places where things are made, not researched, planned, designed, or engineered.

But things worked out differently in Japan. Late development, turbulent markets, and an extreme emphasis on quality, reliability, and time to market for a wide variety of products have made Knowledge Works into functional equivalents of Western, M-Form firms. Thus, separation and differentiation of corporate, divisional, and operational activities have not proceeded nearly as far in Japan as they have in Western Europe and North America. Corporate and divisional level activities are not nearly as distinct from factory level operations as a result.

So, from a functional analogue perspective, Knowledge Works function like minidivisions consolidating almost everything but basic research activities for families of related products in extremely robust and capable factory sites. Somewhat less obviously, such sites could be considered analogous to Honda's strong product development organizations and Boeing's 777 design-build teams, although in neither case are the industrial R & D, engineering, and management skills concentrated at Honda and Boeing factories as rich, versatile, and flexible as those found at Knowledge Works sites.[36] Recall that Yanagicho is responsible for 13 different product lines embodying an impressive array of heterogeneous technologies, people, products, and processes.[37] By combining so many resources so low in the hierarchy, factories become multifunction, multiproduct, and multifocal manufacturing sites, replete with resources and primed with operations and systems integration know-how. In high-tech, fast-to-market competition in Japan, concentrating such resources and managing such capabilities anywhere other than at the level of factories results in too little, too late to market, at too great a cost.[38]

THREE

Organizational Campaigning

Toshiba seeks constant organizational change for the better, and organizational campaigning makes this happen. Organizational campaigning is a kind of programmed, disciplined, recurring activity that integrates a Knowledge Works' multiple functions, products, and purposes, and orchestrates its managers, workers, and engineers, diffusing and amplifying their learning. Organizational campaigning joins factory-based QC (quality control), TQM (total quality management), JIT (just-in-time), and TP (total productivity) methods to enhance profits, reduce cycle time, and more efficiently mesh factory resources with corporate strategy-making. Effectively linking multiple goals and levels in something approaching real time is an organizational achievement of enormous significance. Toshiba would be less competitive and Knowledge Works less creative without campaigning.

Organizational Knowledge and Renewal

The knowledge, capability, and ambition that power Yanagicho are mostly generated from within, through internal processes of research, design, development, and shop floor experimentation that are the factory's most strategic resources. Internal activities have to be clearly distinguished from market forces that are also important, but knowledge, capability, and ambition are contained within the factory and, thus, are more under its organizational control. Full and coherent use of such resources gives Yanagicho a kind of strategic integrity that is almost palpable.[1]

At an earlier time, indeed for much of the 20th century, knowledge was not an internal, organizational resource. Japan imported most of its tangible knowledge from abroad, and so generating knowledge from within was not very important. Today, however, Japanese firms are net exporters of technology, and revenues from the international sale of technology exceed purchases by a considerable margin.[2] Knowledge creation and its application are now the activities most directly responsible for Toshiba's bottom line.

This is good because, unlike Western firms that hire and fire according to demand or buy and sell company parts according to need, large firms in Japan hire most employees directly upon graduation, rarely fire or lay them off, and almost never engage in hostile merger and acquisition activities. Since the ways that firms can exchange, transfer, and transform assets in the marketplace are limited, full and frequent use of available resources are emphasized instead. The drill is to recycle and adapt rather than to lop-off and acquire new resources.

Because market options with respect to asset valuation and conversion are limited, *internal standards and processes* of resource expenditure are crucial. Issues of organizational control loom large as a result. These are especially important for Toshiba, given its long reliance on outside partners as sources of foreign technology transfer. Foreign technologies are no longer readily available nor such a bargain anymore. Reusing and renewing corporate resources, rather than replacing them, are today's goals and, thus, adaptive rather than allocative efficiencies are pushing Toshiba ahead. Organizational campaigning is a primary means for doing so.[3]

The Iron Cage and Campaigning

Big firms, like bureaucratic structures anywhere, often become complacent and resistant to change. The tendency to rigidity and resistance to change has been called the "Iron Cage" of organizations, meaning that they necessarily become more bureaucratic and suboptimizing in the long run.[4] Bureaucratism may be even more pronounced in Japan, given the smaller size and greater specialization of firms there, so less slack is available to offset its forces. Policies of long-term employment and seniority-based compensation may also accentuate a structural tendency to favor bureaucratism.[5]

The kinds of change that are pursued may affect resistance to change as well, and in this too Japan's firms may be different. Small changes are more easily digested than large; voluntary changes more willingly accepted than those imposed by fiat. So, embracing many voluntarily adopted, minor changes, more or less continuously, may be the best way of encouraging large-scale organizational change. The PDCA (plan, do, check, action) cycle of QC and TQM practices do exactly that by mandating teams to manage, measure, and monitor their daily work routines. Given how widely and effectively such practices are diffused in Japan, they counterbalance some of the structural forces of bureaucratism that are all too evident.

QC and TQM are exercised in the interest of promoting employee participation and productivity by encouraging constant yet incremental, or what is known as *kaizen*, change. But there can be too much of anything, even a good thing like *kaizen*. Incessant *kaizen* practice can entice employees to change simply for the sake of change, quickly leading to a point of diminishing marginal returns. All of us have experienced that "here we go again" feeling when a parent, teacher, or boss starts in on us to point out what we had better or better not do. That "here we go again" feeling of diminishing marginal returns, called *mannerika* in Japanese (thoughtless action or apparent change without real change) may be reached more quickly in Japan because *kaizen*, QC, and TQM have been so well practiced for so long. Employees are accustomed to, some would say inured to, organizational change.

Mannerika is a species of suboptimizing behavior related to the timing and repetition of change routines. "Here we go again" really means "here comes the same old thing." Partly because *kaizen*, QC, TQM, JIT, and other change practices are so well known in Japan, employees can pretend to be fully involved but not actually be. Careful attention to changing effort levels and outcomes is needed in such instances, and these should be measured and monitored as functions of time.

Amitai Etzioni, a Columbia University sociologist, developed a model of large-scale organizational change based on what he called cycles of compliance. Compliance refers to the conforming or nonconforming behavior of those who are in the midst of organizational and behavioral change, measured against the expectations and performance goals of those in charge of planning and managing change. Etzioni argued that large-scale change activities move through a predictable sequence of four phases: education and promotion; commitment; performance; decline and withdrawal.[6] The sequences may be seen in Figure 3.1.

While in a general sense the direction of change is predictable, the duration of phases and, therefore, the timing with which phases succeed one another, is less predictable. Also, the degrees of overlap and interpenetration of phases and cycles are not predetermined, so assessing how thoroughly and well change is being accomplished remains a key management task. Avoiding *mannerika* depends on timing. Moving a majority of employees in the direction of change for the better means managing with clear-cut notions of how behaviors change over time.

At Toshiba, what are called campaign (*undo*) cycles—the analogy to Etzioni's compliance cycles—typically last for two years. That is, a campaign can be expected to move through Etzioni's four phases in two years, although a recycled campaign (as opposed to a new one) may move more briskly through the phases. Phase timing and succession depend in large measure on the degree of commitment

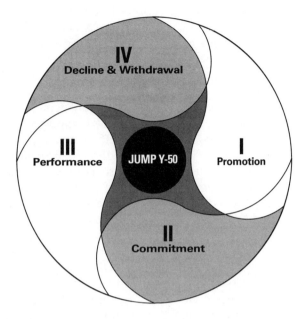

Figure 3.1   A model of organizational campaigning: Yanagicho's JUMP Y-50 Campaign.

and training of campaign leaders who lead, cajole, and move people
through various phases and cycles. Leaders, in turn, seek to mobilize
and train followers, and so on it goes.

Joseph Juran, a godfather of the QC movement, pointed out why
quality control activities succeeded so well in Japan. Juran wrote
that Japanese senior executives took quality control seriously; companies trained engineers in statistical quality control techniques;
companies enlarged their business planning in order to include
quality as a central feature of corporate strategy.[7] Everyone, from
top executives to workers, was involved in QC activities.

The same thing can be said for organizational campaigning. For
an entire organization to change, commitment and training must
be all-encompassing. Even the most remote employees have to be
educated and engaged to be part of the change process. Note that
they do not have to be 100 percent committed, only educated and
engaged. Peer pressure, good management, and the force of a well-
executed campaign will do the rest.

TP Campaigning: From Mie to Yanagicho

Nothing changes without reason. The 1970s and '80s brought Japan
many reasons to change. Spurred in nearly equal parts by the oil

crises and resulting worldwide, economic recession, by the dramatic revaluation of the Japanese yen against the American dollar, and by rising competition from Newly Industrializing Countries (NICs), Southeast Asia, and coastal China, Toshiba, along with a considerable number of Japan's other large firms, initiated TP (Total Productivity) campaigning.

Through the good offices of the Japan Management Association and the Japan Productivity Association, TP campaigning is becoming the single most important tool for affecting large-scale organizational change in Japan. To calibrate individual firm efforts against what was happening more generally, procedures for awarding national TP prizes were adopted in May 1982 under the leadership of Professor Makio Akiba of Tokyo Rika University.[8] TP Management is to the 1980s and '90s what Total Quality Management was to 1960s and '70s: a national movement to improve management effectiveness and firm performance.

Toshiba's own corporatewide productivity efforts began under the direction of Messrs. Sato Fumio and Yamamoto Tetsuya. In June 1985, Sato, an executive vice president of the company, described total productivity as "the creation of a market-oriented ethos embracing all of us throughout Toshiba."[9] The fulcrum for doing so was Yamamoto's recognition that factories are cost centers while divisions are profit centers. Therein lies the rub. A general manager in the Productivity Division at the time, Yamamoto had actually developed the TP concept and program from 1978 to 1982 when he was the factory manager at the Mie Works, near Nagoya. In order to reduce inventory, cut product development lead times, save space, improve productivity, and most important, boost morale, Yamamoto reorganized not just the shop floor but the entire factory.

The problem at the Mie Works, an electric motor factory, was that market demand for so-called standard motors dropped drastically following the two oil shocks. Until then 70 percent of the production mix had been standard units. But the worldwide energy crisis meant that no one wanted standard motors any longer. Customers wanted smaller, lighter, more durable, and more energy-efficient motors. The value of Mie's production fell from $300 to $127.5 million between 1972 and 1978.[10]

By 1980, Mie's production mix had been reversed to 70 percent custom-order goods and 30 percent standard items. A custom order is a batch lot of between 1 and 20 units. In two years, Mie had greatly revamped its traditional product lineup, and thereby delivered the factory from a fate of inevitable layoffs and a possible closing. Layoffs and plant closings were not unknown during the darkest days of the late 1970s, even in Japan.

Thus, the prototype for Toshiba's TP Campaigns was forged in the midst of Mie's struggles. At first, the effort was more trial-and-error than anything else. It is worth noting that 10 months after Ya-mamoto plunged into a full-scale reorganization of the Mie Works, *Toyota no Genba Kanri (Toyota's Factory Management)* appeared in Oc-tober 1978. This book, which describes the efforts of Toyota Motor to cope with issues of factory management and organization in the post–oil-shock world, convinced Yamamoto and his team that they were on the right track.[11] Obviously, what Toshiba and Toyota were both trying to do were not isolated efforts; everywhere firms were re-sponding to turbulent technical, financial, and market conditions.

Dandori tanshuku, or the reduction of setup times in switching over from making one product to another, is the key to small-lot manufacturing. The concept may be simple but the execution is not. Jigs, molds, and dies may weigh hundreds of pounds and are often awkward, bulky, and slippery with machine oil and metal shavings. The changeover process from one setup to another may be difficult, demanding, and sometimes even a little dangerous. Just-used jigs and molds cannot really be discarded or disposed of because they may be needed as soon as the transfer line switches back again to making previous products. Yet, you cannot really leave them in harm's way. So, changeover issues are exacerbated by issues of safety, storage, and frequency of use.

Plus, mixed-model (product) manufacturing may require another set of jigs, molds, and dies every time a new model is introduced on the line. While single-product, mixed-lot manufacturing is compli-cated enough, laymen rarely appreciate the combinatorial com-plexity of mixed-product, mixed-lot production. To Toshiba's credit, all of its contemporary factories practice mixed-product, mixed-model, mixed-lot manufacturing.

When Yamamoto's team first studied the changeover times for Mie's 830 machine tools and transfer machines, they discovered that 79 had setup times of more than eight hours, and there were no machine tools with changeover times of less than one hour. Today, the goal for all machine tools and transfer machines is "single *dandori,*" that is changeover in less than 10 minutes, and the num-ber of "flash *dandori*" changeovers in less than 2 minutes is increas-ing. The shift to making many products and models in small lots is illustrated in Figure 3.2.

Mie's problems were not isolated ones. Most of Toshiba's re-sources (and those of other industrial firms) are located in opera-tional facilities, particularly in factories and divisional (Works) laboratories. Employees at the lowest levels of firms often fail to see the connections between what they do and how companies make money. TP campaigning tries to make that connection by bringing divisional concerns with making profits down to the

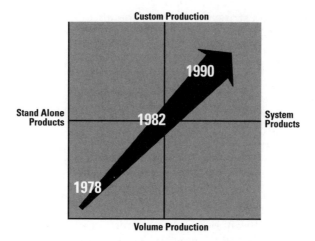

Figure 3.2 TP's transformation of industrial production at the Mie Works, 1978–1990.

shop floor, parts warehouse, and purchasing department. By encouraging employees to focus on profits instead of costs, hopefully they become more involved, participate more, and are more ambitious.

Human nature turns the trick. Convincing employees that the company is theirs is possible by tying pay to profits and by the pay-for-play schemes that reward employees for their suggestions and involvement in how to make things better. Once employees realize that they have a stake in the company and a share of the profits, most willingly participate in TP campaigning. Peer pressure and small team activities help as well. If employees can be made to feel like owners, they can be made to feel like managers in a few more steps.

TP campaigning aims to unify and integrate resources from shop floor to headquarters. It is a grand strategy, mental as much as organizational and social. Yet total productivity is not total quality. It is beyond total quality. Yamamoto came to TP when he realized that most of the gains from QC and TQC activities were already in the bank. Marginal gains were still available, of course, but new gains would be less substantial and important. This is the Iron Cage argument and, thus, according to Yamamoto, something new was needed. TP was new; it differed from previous management practices in the ways shown in Table 3.1.

Even though the impact of Toyota Production System-like techniques and methods was dramatic at Mie, halving inventories and doubling productivity from 1980 to 1985, such gains were dimin-

Table 3.1 Old and New Management at Toshiba

Old	New
mental aspects of work not made explicit and not tied directly to work routines	all corporate resources, mental and physical, made clear and included in productivity planning
difficult to compare goals and targets of different departments	goals and targets of each department clearly developed in context of corporatewide planning
target setting mostly a bottom-up process	goals and targets set at top and passed down for clarification and implementation
through small group activities, like QC circles, efforts to improve productivity at the shop floor	directives and goals set at top even while bottom-up originated processes are crucial to enactment
because individual and departmental efforts were planning targets, productivity management could deteriorate into measuring progress toward piecemeal goals	by setting supraordinate, uniformly high goals, progress is measured incrementally; continuous improvement rather than continuous measurement.

ishing and production was only one part of a long value chain stretching back to purchasing and ahead to point-of-sales information processing, cost accounting, and credit collections. Yamamoto wanted to capture cost savings all along the value chain. That was the genesis of TP campaigning at Mie in 1977. TP campaigning for Toshiba was unveiled in 1985, and by February 1991, the company had more than 300 employees working full-time on TP issues in its 27 domestic factories.[12]

From Factory to Firm and Back Again

Before Toshiba can hope to motivate, move, and manage all of its employees, however, several matters require attention. The first of these is the industrial relations system. Obviously, union power and prerogatives affect the coordination of some 75,000 employees. Less obviously but no less importantly, the cultural nature of the workplace also affects TP planning and performance. Effective campaigning, union disposition, and vibrant local cultures go hand in hand. One depends on the other two.

Industrial Relations Background

Broadly speaking, postwar industrial relations in Japan may be divided into three periods: 1945–49; 1950–55; 1956–90s. The first of these is often termed the *Sanbetsu* period after the left-leaning labor union that dominated labor relations activities at the time. Unions enjoyed an upper-hand in labor negotiations because firms had to obtain their prior approval to dismiss employees or to transfer employees within companies.

Sanbetsu's upper-hand fell after the "reverse course" when the Occupation government's policy toward Sanbetsu and its more radical Communist members changed dramatically following a threatened general strike in 1948. Thereafter, during a shakeout period of some 18 months, a new labor coalition and a new labor bargaining position emerged. The *Sohyo* period from 1950 to 1955, named after the dominant General Federation of Labor, was marked by the so-called Takano line, named for Minoru Takano, general secretary of Sohyo. The Takano line was a do-little line. Because labor and management did not agree on terms for a new labor contract, the Takano line simply hoped for a carry-over effect from the earlier, union-dominated period. Do something, in fact, and union power may be weakened.[13]

Beginning in 1956 and continuing until today, a new era of labor negotiations unfolded. Under this regime, companies may dismiss workers for various reasons, but most notably for not working or for lack of proper behavior at work. Negotiations surrounding these issues were conducted from 1956 to 1965 within a framework of a collective bargaining agreement (the Ohta/Yuwai line) but since then, mere consultations rather than collective bargaining determine terms of dismissal and issues of work content, pace, and rotation.

In effect, during a half-century of extraordinary economic growth, independent union power—independent of large firms—has become progressively weakened. But, unions, along with large firms are credited with providing the higher wages, better benefits, and higher standards of living that have materialized for regular employees since the 1960s. Today, all employees of large firms below the rank of department manager are union members. One union represents all employees in wage negotiations and bargaining, although there may be different branches of the union in various subunits of the firm. In sum, union goals of better wages and benefits have been institutionalized in a system of employment where union activities are firmly embedded in corporate structures and strategies. Hence, union demands for an independent voice in the design and content of work processes have diminished to a point of insignificance.[14] But, at the same time, wages that reflect labor productivity increases are now an accepted part of compensation within large firms.

Labor unions, in other words, are both partners and advocates. They are partners with firms in that practically everyone at Toshiba, for example, belongs to the same union, that is, the company union. Union members and company employees are nearly synonymous. The only difference is that employee welfare and safety are first and foremost in the minds of union leaders whereas top and middle managers are more concerned with profits and returns on investment. But union leaders know that the health of the firm is the wealth of the union, and executives know that the wealth of the firm is the health of its employees.

At the factory level of organization, the role of the labor union at Yanagicho is suggestive of both the problems and the advantages of Toshiba's management system. The union has a very elaborate structure reaching to the shop floor and is in constant contact with workers in departments and sections through frequent meetings and informal discussions between union stewards and workers. In a factory where worker skill and adaptability are in such demand, the potential for overstressing employees and for disrupting their work lives is great. The union walks a veritable tight rope of balancing the interests of the company and those of workers. Those difficulties are well reflected in the words of Mr. Yasuba Kanesaki, speaking as president of Toshiba's Yanagicho Workers Union (Denki Roren).[15] He suggests that the success of TP campaigning, even in Japan, is contingent on the industrial relations context.

Not a Rose Garden

Kanesaki is critical of many things that TP campaigning sets out to do and how it goes about doing them. But in 1987, TP campaigning was only two years old at the time of these interviews, when Kanesaki was 48. He had been factory union president for three years, and general secretary (shokicho) for eight years before that. But by his own admission, he was and is an engineer, previously involved with designing the air conditioning units for the high-speed bullet trains that opened rapid rail service between Tokyo and Osaka in 1964. He misses that work. Kanesaki says he wanted to spend his life as an engineer, but that union work found him.

Kanesaki wonders if he quit now or if he failed to be reelected (as president), where he would go. His previous workplace—the air conditioning and refrigeration department—is no longer at Yanagicho. It was moved to Shikoku. He laughs while he talks about this, obviously pleased with his current work and enjoying it even though he misses his life as an engineer. He is very enthusiastic about getting the opinion of those on the shop floor, especially *seizo*, or production workers, into the dialogue for change. This is exactly what TP campaigning is designed to do.

It isn't that there are necessarily fewer problems in larger companies, he says, but rather that people don't recognize problems as problems. The size of the firm ensures security of employment, and people just say, "size covers a multitude of problems," and let problems go. This is the dilemma of large firms.

"At this factory we do things together—the designers and production workers all work on new products together. From the standpoint of workers this is sometimes a disadvantage. You don't know exactly who is responsible for what. If there is a problem, no one says clearly and assertively, 'You just can't make it this way.' Because designers [sekkeisha] are there, production workers just can't say what they really think.

"*Sekkeisha* have all the logic and theories, and when faced with a concerted defense by designers of a design issue, workers just clam up ["iroiro iwareru to ienaku naru"]. A solution to overcoming this dilemma is to have just a few who are really good at design go to the shop floor, then the production workers won't be afraid to speak up. This is what I did. I was an engineer and I went to the line. I was always criticizing the designs, but I could defend myself on the spot. I caused a lot of trouble [he laughs]. Production responsibility should lie with the line, but *seizo* people are reluctant to speak up around engineers and designers. So, over the long run, operational integration causes problems. It's better if design remains design, and production stays production.

"When an idea is ventured, some of the department heads [bucho] will ask 'who said that, whose opinion is it?' When I hear that, I get angry and tell them, 'It doesn't matter whose it is, if the opinion is right, it's right. It doesn't matter who came up with it.' This kind of authoritarian attitude makes it hard for people to speak up and say what they think. Take *kaizen* for example. The suggestion system movement is popular. *But* . . . the real problem is the quality of the suggestion. People just get into a competition for numbers and rewards and say anything that comes into their heads. There isn't a penetration to more serious issues about work process and product design.

"How can we do so? Leadership from the top [ue no shido]—the union top. That is about keeping promises and accepting responsibility. For instance, keeping promises takes time. You have to start on time and end on time. When we have meetings and some leaders stroll in late and start talking about something or criticizing what we've been working on, I tell them that if they can't come on time, not to bother to come at all or at least not to tear apart what we've been doing.

"What is the most important role of the union now? It isn't working conditions anymore. Now the big issue is protection of jobs. We fire people for things like not working for a couple of weeks without a valid excuse, or even not working hard enough. Those types might as well die, they'd be doing their families and society a favor—not to mention the company. The real problem is that there is too much of the 'take it easy, everything's alright' [maa maa ii jya nai desu ka] type of attitude. This just won't do. But as people get lazier and society gets more complicated, we lose a sense of clarity and start letting things slide. This is the root of many problems. What you really need is people who can argue clearly back and forth. But now, everyone is in the same position, and they're afraid—maybe I won't be liked, maybe I'll lose friends, etc.

"The first signs of this sort of 'I'm OK, you're OK' attitude are seen in the reluctance to greet others [aisatsu] correctly. People don't say 'good morning,' etc. They don't bow to each other like they should. In the training center they do it, and are told to do so if they don't, but once they are in the workplace, they don't have to and often don't. This is where the evaluation system comes in. From the union's point of view, without a proper appraisal system the company can't be competitive. Proper means fair. In general, only one-third of those eligible for promotion at the middle and upper levels of the factory are in fact promoted.

"The changing composition of union membership is a headache too. There are 34 branches of the union in Toshiba. If we tried to apportion work equitably according to union seniority, we'd have real problems on our hands. Members have to understand that the nature of work is changing. Research, development, and design work are being done increasingly by university graduates. The union has to adapt or the company won't be competitive. Factory work is changing. It used to be how many can be made by when [nanjikan ni nandai] but now it's how long will it take to develop it [kaihatsu itsu made]. The mass production age is over [Tairyo seisan no jidai ga owatta]."

Yanagicho and Its Culture

Kanesaki's remarks are revealing. While the specific content of campaigning may be important to increasing flexibility and cohesion among employees in different parts of the factory, it is more the process that links specific shop floor concerns to those of the entire factory. The process builds a flexible and integrated workforce. And process is something that is extremely difficult to manage.

Yet Yanagicho has a lot of experience trying to link parts to the whole and the whole with the parts. It celebrated 60 years of operations in 1995 as Toshiba's third oldest factory. Yanagicho has survived and prospered by constantly reforming, restructuring, and reinventing itself according to the needs of the market, the directives of top management, and most importantly, the ambitions of its workforce. But it was not always so. Before the Korean War, factories like Yanagicho operated largely on their own without much central direction or control. However, as firm structures became more elaborated with a concomitant growth in the numbers and quality of managers in the postwar period, the autonomy of local factories was reduced.

Structurally speaking, a differentiation of control and coordination structures and a roll-out of organizational campaigns have fairly well eliminated the piecemeal and localized quality of prewar factory management. Yet factories are still in the frontline for promoting large-scale organizational change because industrial resources and capabilities are concentrated there and because, in time-based competition, factories are ultimately responsible for making the right products at the right times in the right amounts.

All of this responsibility creates a kind of cultural context. Generally speaking, factories are not credited with having cultures of their own, distinct from company, ethnic, and national cultures. That Yanagicho, one of Toshiba's 27 domestic factories, has its own culture—its own worldview, values, and ethos or its very own way of seeing, acting, and feeling in the world—is saying a great deal. But beyond any doubt, Yanagicho has its own distinctive, location-specific culture. Indeed, Yanagicho has many cultures, since people make culture and there are some 3-4,000 people at Yanagicho. Their many cultures, really subcultures, nonetheless share a history of meanings, structures, events, and group learning processes delimited by factory boundaries, those working there, and the products being made there. Yet it is not a world onto itself. It is part of Toshiba, located in Kawasaki City, and subject to Tokyo Effects.

The Yanagicho culture, to carry the argument further, is the very reason why Yanagicho can campaign organizationally. Campaigning takes these things, relationships, and meanings, and transforms them by selecting, training, socializing, and motivating people as if they were Yanagicho employees first and foremost. While some employees may never internalize such efforts, most do in the long run. Low turnover and transfer rates have something to do with this. Except for personnel department and accounting managers, Yanagicho staffers typically stay there for the greater part of their working careers. This is as true for line workers hired directly by Yanagicho as it is for university graduates assigned there.[16] So while Toshiba employs them, their identities and careers belong to Yanagicho.

The thrust of identification, commitment, and progress are powerful because Yanagicho employees are a relatively small group, some 2,758 regular employees, organized into dozens of subunits. About one-third of these are line workers, and so two-thirds are engaged in managing, planning, coordinating, and analyzing the labors of a modern factory workforce. The strength and integration of the group, as a group, is intensified by the relatively small numbers involved, their relatively high levels of education, and Yanagicho's investment in their socialization and on-the-job training. (Since the 1980s, everyone hired has at least a high school education.)

Small numbers, high education levels, and in-company training are reinforced by creating circumstances that emphasize what employees can do for Yanagicho and not vice versa. QC circle competitions, factory festivals, athletic events, and even the uniforms that everyone wears are examples of how employees are encouraged— really, asked and expected—to contribute and fit in. Almost everything that happens at Yanagicho is a part of a process of cultural sense-making that reinforces a commitment to place, people, and industrial purpose.

Campaigning is part of the common experience of sense-making. Work teams are the means to implement motivational campaigns, making them individually relevant and socially effective. In this way, Yanagicho makes a culture, has a culture, and is always in the process of adding to and changing that culture. Yet this culture is not a loose concept incorporating anything that smacks of values, beliefs, causes, norms, or practices. Neither is it a figment of management's imagination nor a sop to workers' whining. It is an experience of creating, sharing, and molding meaningful institutional and personal experiences that become all the more cogent by acting together.

And it is about learning: the accumulated knowledge of a factory, embodied in and embraced by the people working there.[17] Not everything in this reservoir of learning is immediately available to managers and workers, however. Deep-seated feelings and values cannot simply be called up and made part of a constant process of cultural negotiation. Tacit knowledge is just that, tacit. And tacit knowledge is a big part of the Yanagicho experience. Little can be accomplished without taking the entire reservoir of experience and knowledge into account. Campaigning tries to do this by challenging employees to reconsider how and why things are done the way that they are. Yanagicho workers feel this. When I interviewed the head of one of Toshiba's overseas plants and his right-hand man in 1990, one of them told me dispiritedly, "There is no culture here" [koko niwa bunka ga nai]." The other retorted, "We'll make one from now on [kore kara tsukuru]."[18] (See chapter 6 for more on overseas and domestic factories.)

Yanagicho and TP Campaigning

While the history of TP campaigning stretches back to Yamamoto's stewardship of the Mie Works, TP campaigning is not Yanagicho's first campaign. Far from it. A "Creating New Values" campaign preceded it, and a "Y-Jump 50" campaign came before that. More preceded these. The "Y-Jump 50" campaign, which celebrated Yanagicho's 50 years of operations, for example, appeared in the mid-1980s as a campaign to increase productivity, efficiency, quality, and profits. Yanagicho was not performing poorly in these areas but management's task is to anticipate problems before they result in poor performance. Yanagicho's 60-year anniversary provided yet another occasion for doing so.

The "Y-Jump 50" campaign could not have come at a better time because Yanagicho, Toshiba, and the national economy were doing extremely well. No one expected that in a few short years, the economy would slip into a tailspin and Toshiba's astonishing success with microchip and IC device production would flatten out. The "Y-Jump 50" campaign's anticipation of such problems was not accidental or coincidental, however. The real task of management, according to Toshiba lore, is not claiming credit when things are going well but avoiding periods of crisis when things go bad. A Japanese proverb to this effect admonishes warriors to tighten their chin straps when victory is at hand (nodo wo o shimeru).

The "Creating New Values" campaign was a natural outgrowth, of the "Y-Jump 50" campaign. However, what was different about the "Creating New Values" campaign was its emphasis on human resource (HR) training. Not that human resources were unimportant previously, but HR training was not so consciously and conscientiously tied to the goals of efficiency, productivity, quality, and profitability. "Creating New Values" made those connections and assumed that further advances with respect to quantitative goals, like productivity and profitability, would require a qualitative change of mind. "Creating New Values" had seven subthemes: consumer orientation, factory modernization, greater speed and better quality of work, heightened value-added manufacturing, heightened competitiveness, greater self-awareness, stronger people and stronger culture.

Obviously, the first half of these relate to the "Y-Jump 50's" emphasis on quantitative progress while the latter half are more concerned with subjective goals, like self-awareness and strong culture. This underscores the importance of continuity and learning in making participants comfortable with the organizational campaigning process and ensuring that new endeavors are connected with past efforts. Unless such connections are made, it is all too easy for employees to be mindlessly carried along (*mannerika suru*) by campaigns.

Subjective awareness and commitment were themselves campaign goals in the "Creating New Values" campaign, and such personal and subjective goals are what laid the foundation for the TP campaigns to follow. By 1995, 10 years after "Y-Jump 50," TP campaigning was already in its fourth cycle in some Yanagicho departments. The first cycle, in the late 1980s, was connected to TQM and *kaizen* themes, followed at the turn of the decade with an extension of *kaizen* techniques to design and sales engineers, groups that have been on the periphery of TQM activities. Next came a strong push in high-volume production areas, like photocopiers and laser beam printers, where the goal was to make just enough to satisfy demand. No more, no less.

"No more, no less" was accomplished by a tactic of "focused flexibility." Departmental resources were focused on Product A, until the demand for Product A was satisfied. Then, the focus shifted to Product B, Product C, and so on. The fourth TP cycle, commencing in late 1991 for high-volume departments, aimed at mixing production for models A, B, C, and so on. Mixed model production is actually a series of small batch runs; batches of as few as five units come down the line together. Balancing all the needed resources—human, technical, organizational, and managerial—in mixed model manufacturing requires a virtuoso, high-wire performance (see the discussion of the Champion Line and Balanced Line models in the previous chapter).

Balancing is made possible by a careful rollout of campaign targets and goals at the factory level of organization. During the initial phases of any campaign cycle, there is a closely monitored minicycle of PDCA (planning, doing, checking, and acting) whereby every department and section that will be involved in the next phase of campaigning sets its goals, has them checked for consistency and integrity, has them legitimated by higher-level authority, and then moves ahead with them. This minicycle, as illustrated in Figure 3.3, is a key activity for ensuring the overall continuity, coherence, and integrity of campaigns.

Virtuosity in campaigning is possible only when a strong foundation has been laid, practice is constant, and ambitious people are

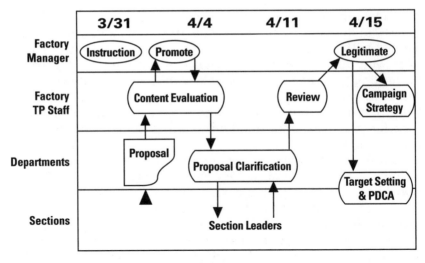

Figure 3.3   TP target setting schedule.

involved. Indeed, not every product department in Yanagicho manages to realize virtuoso performances and advance to higher cycles
or levels of TP campaigning. The custom goods side of the factory,
making automatic postal sorting machines or turnpike toll systems,
for example, were still improving the functional integration of design and sales engineering in 1991, cycle two of the TP campaign,
while the volume production side of Yanagicho had moved past
cycle three.[19]

Negotiating Cultural and Organizational Fit

Anthropologists refer to the sense-making of individuals in organizations as a social negotiation of individual commitment.[20] It is unlikely that anyone at Yanagicho or any other organization for that
matter is sufficiently well read in cultural theory to be designing organizational campaigns according to some notion of how social and
individual commitment are negotiated. However, given the consistency with which campaign goals and schedules are articulated,
there can be little doubt that those planning campaigns understand
the subtleties of how they work. The aim is to design an interactive
process whereby all employees, no matter their level or functional
specialization, think through how their work relates to the whole.
The PDCA (plan, do, check, action) cycle of TQM and TP activities
forces employees to do exactly that, and doing so creates a social
framework of individual choice that attenuates the top-down side of
TP campaigning.

Continuity of campaign goals and methods assumes that there will be progress, even though it will be partial and piecemeal. Continuity in language and goals facilitates dialogue, interaction, and even debate around a set of common themes and goals by using common language and common performance measures. Campaigning is a process of creating a consistent means for affecting organizational change, and in this way "Y-Jump 50" and "Creating New Values" were the groundwork for TP campaigning.

Mobilizing, Motivating, and Moving Employees

TP campaigning requires firms to motivate enough employees to buy into the change process without appearing to control or manipulate the process from above. This is no mean feat when firms employ tens of thousands and there is surprisingly little in common to unite and motivate employees across many divisions. The trick, if there is one, is found in phase one of the compliance cycle: enough employees have to be educated and willing to change their own behavior, assuming, of course, that both long-term processes of employee socialization and training, and short-term connections between firm profitability, factory productivity, and individual performance are in place.

Thus a clear distinction must be made between company efforts to control employees and company campaigns to motivate, engage, and help them create their own workplace norms and values. TP campaigns are mostly about the latter and are effective because of the prior socialization and training of employees who have already experienced the "Y-Jump 50" and "Creating New Values" campaigns. Also, the interlocking features of the Japanese employment system—lifetime employment, seniority based compensation, and enterprise unionism—provide a basic framework within which the political and economic interests of firms and employees converge more than diverge.

The framework and context of work in Toshiba are distinctive, based as they are on factory-specific cultures and workplace-embedded techniques of organizational campaigning.[21] While Yanagicho management wants workers to be aware of these differences, it also wants to minimize the differences. They are minimized by the thoroughness with which campaigns capture and engage employees. Although some employees may slacken their efforts and dither along, such behavior engenders a strong resistance to the meanings of employment at Yanagicho and TP campaigning. And the TP effort is not isolated. It succeeds and complements other campaigns, and it is part of a wider factory strategy for becoming more competitive, flexible, and responsive. Indeed, in 1991, TP campaigning was only one of three campaigns mentioned in Yanagicho's semiannual business plan, and taken together, those campaigns were only one of seven goals in the plan:

1. Realize the factory's profit objectives.
2. Secure the quality, cost, and delivery targets for new products under development.
3. Remodel (and strengthen) the factory's production capabilities.
4. Strengthen the business foundations of the factory through the Y-TP, Y-STEP, and Y-CIM campaigns.
5. Firmly establish new product departments.
6. Refine our administrative routines and capabilities.
7. Create a profoundly safe work environment.

Looks can be deceiving, however. The TP Campaign originated in Toshiba's corporate headquarters, and at Yanagicho, the factory's General Affairs Department (*soomubu*) was heavily involved. The weight of these two units behind the effort bestow considerable authority and legitimacy. At Yanagicho, there are three full-time managers, all industrial engineers, responsible for the TP rollout. They meet frequently with Yanagicho's 20 department managers and more than 80 section managers. And, since the TP Campaign, like most other campaigns, will run again after the current cycle is completed, having so much push behind the effort ensures that what does not go well at first is likely to go better in later phases and cycles.

Next we look at TP campaigning on three levels: support department, product department, and factory. By doing so, the dialectic of making everyone responsive and interconnected by creating, negotiating, and operationalizing goals at different levels can be followed.

1986–89 TP Campaigning at Yanagicho

The Machine Tool Department

The Machine Tool Department at Yanagicho was established in 1938 to manufacture utility meters.[22] Throughout the years, the main work of the department has been the design and maintenance of tools, dies, and molds used in the manufacture and assembly of all sorts of Yanagicho products. In the 1950s and '60s, when Yanagicho burgeoned with new consumer products such as washing machines, refrigerators, and air conditions, the department swelled to as many as 240 regular employees. About 200 of these were assigned to traditional tasks of designing and maintaining tools, dies, and molds, but

40 or so were specialized in the design and development of manufacturing tools needed to support Yanagicho's proliferation of products.

Indeed, without a robust Machine Tool Department, Yanagicho could not boast the number of products and variety of models that it does. The task of the department is to develop the tooling for manufacturing products in volume and variety, and it does so by making prototypes based on drawings produced by design engineers. In short, the Machine Tool Department breathes life into every product made and assembled at Yanagicho. Given that certain products, like photocopiers and laser beam printers, are redesigned frequently, as often as every six months in some cases, Yanagicho's rapid pace of product development hinges singularly on the quality of the Machine Tool Department. The department's importance is evident in the supply of foremen who have advanced to product department managers after being seasoned in the work and rhythms of the Machine Tool Department.[23]

In the mid-1970s, as more and more staff functions were subsumed in product departments as a way of speeding up development and design cycles, the Machine Tool Department shrank to about 150 regular employees.[24] However, if anything, the importance of the department as the pivotal interface between design conceptualization and prototype development grew. As product life-cycles shortened, especially for the volume manufacturing side of the house, there was relentless pressure to outsource the manufacture of parts and components. Outside suppliers, as discussed in the next chapter, generally boast across-the-board lower costs for labor, land, energy, and transportation. So, managers of high-volume products want to outsource whatever they can. But managers from the custom-order side of Yanagicho, say automatic mail sorting equipment, feel that suppliers are not sufficiently tuned in to their needs for lower-volume but nonetheless high-quality and low-cost parts.

Mediation of factory needs and supplier capabilities is one of the crucial tasks of the Machine Tool Department. If, for example, four new photocopier models will appear in the next year, designers from the Machine Tool Department will decide which parts and components should be made in-house, calibrate internal capacity against outside supply, and thereby decide what phases of development and production will be handled internally. The Machine Tool Department along with Purchasing, traditionally the department that actually manages suppliers, decide the actual mix of in-house design, development, and manufacturing activities. Without their expert intermediation, Yanagicho's performance would be notably less impressive.

With respect to TP activities, the role of the Machine Tool Department is equally impressive because its assessments of what can and should be developed in-house fits the top-down directions of management with the bottom-up suggestions of line and staff employees. In 1988, for example, the department set a four-year target of reducing lead times for delivering photocopier (PPC) frame side panels and molds to the PPC Department. The pressed metal parts for the side frame had a 20-day lead time in 1987 and by 1990, the target was to achieve a 25 percent reduction in lead time by providing parts within 15 days. An action plan to pursue lead time reductions was developed. Improvements were sought in four areas: the speed and efficiency with which work was accomplished; improvements in the use of machine tools and setup times; productivity; and quality. Nine steps were identified to realize these goals:

1. Reconfigure the layout for parts production by machine tools.
2. Improve the capacity utilization for machine tools.
3. Reduce the time that it takes to move material from one shop to another.
4. Reduce set-up times for machine tools.
5. Redesign material flows for automatic assembly.
6. Verify the reliability of jigs and molds within two test runs.
7. Improve kit marshalling for parts, tools and material.
8. Enhance computer-assisted design, manufacturing, and testing functions.
9. Generally, reduce waste and slack.

There is nothing remarkable about the four areas for improvement or the nine steps for realizing improvement. What is remarkable are the ways in which areas and steps for improvement are tied to specific targets, the realization of which are the responsibilities of named section heads and work teams. Both department and TP campaign managers and section leaders are named. The tying of reasonable goals to well-specified and mutually agreed upon methods, timelines, and responsibilities is the essence of TP campaigning.

Labor Savings Equipment Department

Moving up one level to the Labor Savings Equipment Department or, as it was renamed in 1988, the Special Equipment Department, finds an increase in the breadth, level of detail, and complexity of TP campaigning. Also, it offers another level of concreteness with respect to the short-term effects of TP campaigning because campaign data exist from 1986, the first full year of TP campaigning. As seen below, TP planning targeted three areas: concrete measures of over- all performance; action plans with respect to particular products and processes; managers who were responsible for aspects of campaign promotion.

A year and a half of performance measures or three half-year to- tals are reported in five categories: manufacturing inventory, value- added manufacturing, setup times, lead times, JIT parts delivery. While the data do not represent a long time series, figures are pro- jected to the end of the 1989 fiscal year (ending in March), provid- ing seven semiannual measures of TP progress. A reduction in lead times is one of the most remarkable areas of projected improve- ment, dropping from 120 days on average to just 58 days. During the three periods where actual data are known, the reduction was an impressive 50 days. Given that the Special Equipment Department operates on the basis of market orders, rather than on the basis of projected sales, a reduction in lead time is especially important for keeping customers happy.[25]

The percentage of single *dandori*, or mold and die changeovers, in under 10 minutes likewise climbed impressively. In the latter half of 1986, the percentage was 90 percent, whereas three years later, in the latter half of 1989, the figure stood at 98 percent. Putting dollar values on such productivity enhancement efforts is a good way to make obvious the benefits of TP campaigning. When the Special Equipment Department did so in the late 1980s, the value of labor (calculated by the turn-out value of what was produced) moved from $290/hour to $430/hour.

Value-added manufacturing was another area of impressive per- formance, advancing index numbers from 0.9 in the first half of 1986, to 1.75 during the latter half of 1987, and projected to go as high as 2.4 by 1989. Value-added improvements on this order were only possible with the close cooperation of the Machine Tool De- partment, because they were responsible for increasing the level of in-house manufacturing as opposed to outsourcing. Recall that lev- els of in-house manufacturing depend critically on the molds, dies, and specialized tools of the Machine Tool Department.

Factory-wide TP

The efforts of departments are rolled into factorywide activities in a number of ways. Until recently, the 3C Shop Movement was one of the most obvious. This minicampaign was launched in September 1990 for a period of two years. 3C stands for a Clean, Comfortable, and Creative shop floor. The idea was that the Machine Tool Shop and the Financial Equipment Section (making ATMs and cash handling machines) of the Labor Savings Equipment Department would cooperate in creating a "model shop" that was clean, comfortable, and creative, anticipating a human-centered, information-intensive, 21st-century workplace.[26]

Like most campaigns, the 3C movement was designed to move through a two-year cycle (Etzioni's compliance cycle), culminating in new ways of working for all the units involved. In this respect, the 3C movement is analogous to the "New Spirit" and "Creating New Values" campaigns. Minor campaigns are run in tandem with major campaigns, like TP campaigning, in order to pinpoint and solve problems in particular areas or processes and, in all likelihood, to deflect some of the inevitable alienation and *mannerika* that constant TP campaigning may engender.

A big difference between the 3C campaign and TP campaigning, however, is the level of authority that was brought to bear in promoting and implementing campaigns. The 3C campaign was headed by a working group composed of a cross-section of employees from the two units involved, the Machine Tool Department and the Financial Equipment Section of the Labor Savings Equipment Department. Thus, the group was an interdepartmental one, made up of section heads and line employees from both departments.

Nevertheless they decided to do a lot. Six objectives were targeted for the fall of 1990 and the winter of 1991: fix the guidelines and policies for the 3C movement; survey the current situation and identify the areas to be targeted; decide on how things should be changed; roll out a *kaizen* program; set up a "model shop"; implement the 5Ss. The 5Ss are a set of concepts for creating a clean and orderly workplace: They are *seiri* (precision), *seiketsu* (neatness), *seiton* (order), *shisuke* (arranged), and *seiso* (spotless). As one section manager put it, "We want our factory clean enough so that when customers visit, they want to buy our products."[27] The 5Ss are pretty well diffused and practiced among most large industrial firms in Japan.

The planning during the fall of 1990 and the winter of 1991 identified 46 specific targets in 4 general areas for improvement during the 1991–93 period. The targets were cross-referenced with respect to the specific advantages that would accrue from them, such as improving elbow room on the line, noise reduction, odor alleviation, and such. Most important, during the implementation period, 3C planning groups were formed and individuals made responsible for the rollout of different parts of the program. Jobs were tied to a task-specific hierarchy.

To give a sense of how many people are involved in minor campaigns, the numbers for the 3C campaign were as follows: overall planning, 3 managers, 2 of whom came from the machine tool area and 1 from financial equipment; 7 deputy managers, evenly distrib-

Organizational Level	Top Management	Factory Management	Department Management	Section & Work Team	TP Specialist
Design Of Overall Campaign	■				
Rollout Of Specific Goals	●	■	●		
Setting of Appropriate Targets		●	■	■	
Coordination of Campaign Compliance	●	●			■

(row group label: TP-Task)

■ Main Responsibility
● Assisting

Figure 3.4 TP campaigning and the division of labor.

uted from the 2 subunits involved and augmented by specialists in operations management from a staff department; implementation, 3 managers and 6 work team leaders, supported by 3 staffers. In sum, the 3C program had 22 managers involved directly and indirectly in designing, developing, and implementing 3C activities.

Given the level of managerial involvement in the 3C program, a much greater effort at mobilizing support for TP campaigning should not be surprising. And obviously, the two are complementary. Specific areas for improvement in the 3C campaign dovetail nicely with overall TP aims. In fact, the 3C campaign is only one of some 39 different department-based campaigns underway in the last half of 1987.[28] Complementarity in department-specific goals and in TP objectives provides a powerful rationale for running many different campaigns at the same time, as discussed earlier.

A differentiation of department-based activities is an essential complement to TP campaigning. Connecting the targets of individual departments, such as the Software Department's desire to become more cost conscious, the Optical-electrical Disk Drive Department's aim to cut in half the time needed to search for and retrieve specific files, or the IC Card System Department's wish to file the largest number of patents among Yanagicho's 20 departments, to the overall aims of the TP Campaign is a crucial task. The integration and coherence of these many activities are department and section management's responsibility, the responsibility of the Factory Manager, and the TP Planning Group. Farther up the line, higher level activities are the responsibility of higher-level managers. The specific assignment of campaign responsibilities by organizational level is an important feature of TP campaigning, as Figure 3.4 details.

As TP campaigning entered its third corporate cycle in 1990–91, no less a person than Mr. Sato Fumio, then vice president and later president, urged all employees to make even greater efforts to realize the goals of small-batch production, decreased lead times, a simple (lean) production system, and a reduction in process inventories. With these in mind, Yanagicho's TP Planning Group identified general targets for custom-order goods and volume-production goods. For example, for ATMs, a custom-order product, a goal was to reduce the parts count in half; for PPCs the target was to have no more than two weeks' supply of key parts. Holding sales and production coordinating meetings twice a month for custom-order goods, and four times a month for volume products, were other goals. It was felt that more frequent meetings would result in better coordination, shorter lead times, and more streamlined production processes.

In general, as these examples attest, the lower one goes in the organizational hierarchy, the more specific TP targets become. The TP Planning Group meets two or three times a month to ensure that factory and department targets are specific and in synch, and the Planning Group tours the factory once a month to answer questions and assess TP progress. Factory departments, for their part, provide monthly TP progress reports and publicly present the results of their efforts once a quarter. The factory manager occasionally makes the rounds with the TP Planning Group, lending his authority to the TP endeavor.

Departmental targets are specific enough that they may be represented monetarily. In 1991, for example, the Utility Meter Department was expected to save 720,000 yen in the costs of making their products during a six-month period (August 1991 to February 1992). The Automatic Mail Sorting and Bank Note Equipment Department was given a target of 930,000 yen to save during the same period. And so on with other departments.

Other than reducing the manufactured cost of products, TP goals for 1991 at Yanagicho included a strengthening of product development capabilities. Again, specific temporal or monetary targets were identified. For example, software engineers working on a new cash handling machine were assigned monthly targets of 5,000 lines of code. Yanagicho's foreign operations were also included in TP goals. By November 1991, for example, Toshiba Systems France, Yanagicho's PPC production site near La Havre, was to design a plant layout and facilities plan for producing toner overseas.[29] In short, every department has specific goals, targets, and marching orders, making across-the-board, measurable improvement the overarching TP purpose. Hence, while comparatively large numbers of managers are involved in TP campaigning, the numbers of full-time managers are relatively small. In 1990 the staff of full-time TP managers at Yanagi-cho was four with seven more part-timers.

Capabilities and Campaigning

In the first five years of the TP effort, 1985–90, Toshiba claims a 20 percent annual improvement in productivity and a 20 to 30 percent improvement in space utilization.[30] Also, Toshiba says that most products have somewhat fewer parts and require somewhat fewer manufacturing steps now than before the start of TP campaigning, although there are difficulties of calculating in-process manufacturing efficiency given the frequency of product and process innovations.

At Yanagicho, the head of TP activities estimates that the campaign has resulted in about a 5 percent increase in productivity each quarter or 20 percent a year.[31] This corresponds nicely, perhaps too nicely, with the corporate estimates of productivity enhancement noted above. However, at least for the IC Card Department, progress has been real and substantial. From the winter of 1987 to the spring of 1988, the department logged a 29 percent drop in manufacturing inventory, a 38 percent improvement in shortening lead times, a 43 percent advance in realizing development milestones, and a 5 percent improvement in overall manufacturing efficiency.[32]

To make local achievements like those in the IC Card Department more broadly available, all factory managers and department heads receive four days of annual training in the latest TP initiatives, and each factory undergoes a month-long period of focused TP activity every year. The goal is for everyone to understand not only what is being targeted but why. By doing so, Toshiba is dealing with local differences that might block corporatewide acceptance of TP goals and activities.

TP campaigning aims to show people how to work better, rather than simply telling them how to work better. But linking employee behavior to TP targets and measuring outcomes are no mean tasks. How should the costs and benefits of TP campaigning be calculated? What percentage of campaign costs should be allocated to corporate, divisional, and operational budgets? Can sales and profit growth that hinge on TP activities be clearly distinguished from more routine increases in sales and profits? There are no easy answers to such difficult questions.

Asking the questions, however, helps focus our minds on the necessity, cost, and efficiency of campaigning. The Iron Cage justifies the necessity, but efficiency really hinges on the costs of campaigning. And costs have to be weighed against the intent and purposes of campaigning: to shake things up by making people rethink what they are doing and how they are doing it. The ability to shake things up and not worry too much about consequences, including costs, is related to success at managing change as a process, not as a product.

Organizational Change and Campaigning

The real trick is not solving problems but finding them, or so many managers say. Organizational campaigning is a way to find and solve problems associated with bureaucratism, localism, and particularism—ailments that habitually afflict large organizations. It is also a way to experiment and innovate by capturing the richness of employee experience. Ultimately, campaigning forces different and often distant subunits of an organization to pull together, find and forge common purposes in campaign rhythms and goals.

Campaigning socializes and motivates employees by enacting change, monitoring performance, and adjusting the tempo and duration of phases and cycles. Organizational campaigning moves Knowledge Works by identifying problems, trying to solve them, and giving employees a sense of commitment and participation. A positive, if occasionally problematic, link between local goals and overall purpose and profits is the desired result. The firm and its employees are at once connected through organizational processes of social, personal, and technical interaction that produce tighter orchestrations of routines, activities, and rewards. While leadership is important to the process, campaigning is independent of the managers who promote it. TP campaigning is not like championing behavior that pivots on the role of particular managers to advance specific products and markets.[33]

The possibilities of large-scale organizational renewal pivot on the degree to which tactics like TP campaigning can be made routine. The gains realized so far are encouraging and the likelihood of future progress seem assured. TP campaigning is transforming Toshiba and Yanagicho into more efficient, profitable, and committed workplaces. Dr. Shoichiro Saba, former CEO and chairman of Toshiba and now a top advisor to the Keidanren, Japan's principal business council, calls TP activities "management engineering." He uses the term advisedly. TP aims to unite everyone—every division, department, section, and work group within Toshiba.

The scale of the task and a constant need for applied learning warrant the engineering label. The underlying goal of TP activities is to reveal the costs that are no one's responsibility and thus are not thought about. By thinking about these costs and making them transparent, efficiency picks up and productivity advances, or so the

TP story goes. Ultimately, the main aim of TP campaigning is to reduce costs by making products only as fast as they can be sold. In this way, TP campaigning is mostly a managerial affair. Campaign strategy, timing, and intensity are management's responsibility. "The real purposes of TP are to correct the errors of management in the past, and to use the past as a guide to the future," concluded Tetsuya Yamamoto in an August 1991 interview.

But history is only one of several reasons why TP campaigning has assumed such contemporary importance. First, as emphasized earlier, the structure of industrial relations supports constructive collaboration between labor and management, and TP activities have to be understood within a context of positive and cooperative industrial relations. Without the active consent of unions, top-down efforts to change firms in comprehensive ways would go nowhere. Even so, as the words of Yanagicho's local union boss attests, there is a lot of rank-and-file grumbling with the constant efforts to manage change.

Second, wages are tied to productivity performance. A flexible wage system awards employees a large portion of their remuneration on the basis of productivity performance. Three elements go into evaluations: growth in corporate (divisional) productivity; work team productivity; and individual productivity. Also, employees are judged not only according to their results but also their attitudes. Indeed, for calculating wage bonuses, results and attitude are all that count.[34] By tying wages to productivity, firms are ensuring that campaigning is meaningful in the lives (pocketbooks) of most employees.

Moreover, respect for hierarchy and for persons in authority are important prerequisites for organizational campaigning. If persons lower in the organization question the targets and goals of TP campaigning as set by persons higher in the organization, then campaign compliance and performance will be poor. Progress will be minimal, and a positive cycle of goal attainment and reward will not be established. In other words, if managers do not exercise legitimate authority, organizational campaigning will amount to little.

Third, firms have invested in productivity enhancement by spending a lot on new and upgraded plant and equipment. In 1990, *The New York Times* reported that capital investment in Japan came to an astonishing 27 percent of GNP.[35] Investment on this order requires that employees learn how to use newly purchased or upgraded equipment and facilities because, once again, productivity performance is tied to remuneration and promotion, and these to job satisfaction, social standing, and commitment.

Fourth, both on-the-job and off-the-job education and training are designed to enhance employment performance. In 1989 alone, Yanagicho mustered 89,000 person hours of off-the-job training in

eight different training categories.[36] Since on-the-job training hours
are thought to be more numerous and important than off-the-job
training hours, the combined totals are noteworthy. Off-the-job
training averages 30 to 35 hours annually. Since employees do not
receive equal amounts of training every year, those receiving off-
the-job training are getting two to three weeks of training per year,
an impressive investment in human resources.

Also, TP campaigning is not an isolated phenomenon. Within
Yanagicho alone, a half-dozen different campaigns are underway at
any one moment, although the TP campaign is the first one to have
everyone within Toshiba doing the same sort of thing at the same
time.[37] Firms are used to managing change through campaigns.
Within a year of the TP kickoff in 1985, TP topics first appeared as
QC topics. Five years later, in 1990, TP issues were the main subject
of QC presentations throughout Toshiba. In fact, most middle-level
managers at the factory see no differences in TQM and TP activities.
The STEP campaign, for example, that began $1\frac{1}{2}$ years after the TP
campaign, is now tightly linked with TP activities.[38] TP activities are
part of everyday work and management at Yanagicho and Toshiba.

Finally, Yanagicho's site-specific culture, an accumulated, nurtured,
engineered, managed, and learned mixture of experience, invention, and insight, has allowed TP campaigning to gain a local foot-
hold and assume a position of importance within factory life. Campaigning affects not only the structures but also the meanings of
work; the constant emphases on self-improvement and self-discovery
complement the more quantitative parts of campaigning, and these
have become part of general processes of sense-making and work negotiation. A culture has been created in which it is possible to capture the myriad inputs of employees, modify them accordingly, and
see if and how the changes work. Committed workers and a lively
workplace result.

Given the importance of TP campaigning at Toshiba and elsewhere, there has been surprisingly little attention paid to campaigning and its effects. Perhaps this is because of the negative press
that mass social campaigns, like China's Great Leap Forward and
North Korea's Flying Horse (*Ch'ollima*), have attracted elsewhere.
Nevertheless, TP campaigning is hardly ever mentioned in Western
language sources on Japanese management, despite its key role in
corporate strategy-making. Perhaps observers have failed to distinguish between the newer TP activities and older QC and TQM movements. The latter are better known and well studied but they are also
30 years old.

TP campaigning is a way of life, a means of keeping up to date,
and Toshiba's principal tool for encoding a set of routines and skills
for coping with the need for constant organizational renewal. "Man-

agement is TP," says Yamamoto Tetsuya, thereby making organizational campaigning synonymous with management at Toshiba.[39] Yet TP campaigning can be as problematic as it is strategic because it aims to make routine the nonroutine by making knowledge creation and innovation widely practiced and predictable. Nice tricks, if Toshiba can do them.

FOUR

Managing Competition and Cooperation

Supplier Organization and Governance

In competition today, factories like Toshiba's Yanagicho Works are only as good as their suppliers. Indeed, during the 1980s between 73 and 83 percent of the manufactured value of Yanagicho's products was sourced from suppliers.[1] Yanagicho continued to outsource high levels of parts, components, and subassemblies from domestic suppliers despite prolonged merchandise trade surpluses, increased foreign direct investment (FDI), and heightened internationalization of the economy. Yanagicho and Toshiba are caught between a decided dependency on literally hundreds of domestic suppliers for most of the value-added content, quality, and price/performance of their products, and a growing international division of labor that is forcing outsourcing and jobs overseas.

This competitive reality has transformed assembler-supplier relations from traditional notions of asymmetrical exploitation of small firms by large into something more akin to business partnerships and alliances. Recall that for Toshiba's factories, valuable organizational knowledge is an accumulation and refinement of production knowledge and practice. The two are inseparable, yet that interdependence pivots on open, intense, and frequent information exchange with hundreds of suppliers, each of whom are different in terms of how they came to be involved with Toshiba and what they expect to gain from Toshiba. Unless the assembler-supplier interface is permeable and, in some crucial ways, equitable, the generation and application of really valuable organizational knowledge may be problematic.

For Toshiba and Yanagicho, therefore, relations with key suppliers—how manufacturers respond to, join, and combine with core suppliers—are pivotal to manufacturing strategy. This is clearly highlighted by Yanagicho's use of 100,000 different kinds of parts and components in its assembly operations, and the issuing of 60,000–70,000 purchase orders for buying 30,000–35,000 different kinds of parts every month.[2] Given the complexity of Yanagicho's products

92

and production/assembly processes, a robust and versatile system for managing suppliers is required. Clearly, suppliers have to be both good and friendly, but being both good (competitive) and friendly (cooperative) is asking for a lot. In conventional economic thinking, suppliers can be good and thus hard to bargain with, or friendly and under your thumb in some financial or managerial way. And that either/or situation makes it hard to realize both competition and cooperation in assembler-supplier relations.

One of Yanagicho's key innovations in this situation—one that allows Yanagicho to pursue both competition and cooperation in organizing and managing its suppliers—is the establishment, encouragement, and guidance of an effective governance system with core suppliers, and this has been realized by the creation of a Supplier Association as detailed in the next section. While this approach may be working at home (not surprisingly, given all the energy directed into the effort), chapter 6 suggests that it may not be working so well overseas.

Network Organization and Governance

Any alliance implies both autonomy and interdependence; partners decide when and how to share risks and rewards. Today, such partnerships are *not* typically secured by equity holdings in the Japanese electronics industry. Perhaps the industry is too fast moving for that, or perhaps a plethora of small- and medium-sized firms makes equity investments by large firms in small firms unnecessary. Instead, intangible asset specificity or knowledge sharing is increasingly favored.

Because knowledge sharing between assemblers and suppliers is absolutely strategic, it is not entrusted to the invisible hand of the marketplace. Instead, a calculated, forward-looking, decidedly managerial attitude drives decisions of what to share, when to share, and how to share. Because partners are independent and their relations are unsecured by equity investments or Western-style contracts, issues of governance—how to govern and manage interfirm relations—become strategic.

If tangible assets are being shared, such as fixed costs in hardware-based telecommunications and transportation systems, the costs of sharing can be estimated with a high degree of accuracy. Costs are related to economies of scale in investment and frequency of use, and so the costs of network membership, operation, and even disintegration are easily estimated.[3] Governance is not a major issue in such cases.

However, when knowledge and know-how are being shared,

costs and rewards are not easily estimated. In these instances, *the visible hand* of management defines rewards and incentives precisely because transactions are motivated by what is invisible. Knowledge-sharing requires a means of measuring value by something other than standard net present value formula, returns on investment (ROI), or returns on equity (ROE) calculations, coin of the realm when valuing tangible assets. The value of intangible asset sharing is not so easily calculated, especially when the speed and effectiveness with which knowledge can be mobilized are crucial.[4]

Because knowledge sharing is hard to define and even harder to manage, a variety of interfirm governance mechanisms, such as supplier industry associations, supplier grading schemes, exchanges of engineering and managerial personnel, and dedicated training programs like Total Quality Management (TQM) and Total Productivity Management (TPM) for the diffusion of technology and know-how, appear. All of these—associations, schemes, forums, training programs, exchange activities, and other forms of information exchange and knowledge sharing—are designed to help partners organize, manage, coordinate, enhance, and reinforce mutually beneficial trading relations in the absence of clear-cut, market-based means for doing so.

In these circumstances, the importance of governance mechanisms and arrangements is likely to be *directly proportional* to the value of shared assets. That is, the greater the asset specificity and the more important partner relations, the more attention is given to managing the growth and governance of supplier relations. So, as interfirm asset specificity has increased in Japan, most notably during the high-growth decades of the 1960s, 1970s, and 1980s, so has an emphasis on effective, equitable, vital partnership and supplier relations, and on the speed with which such relations can be activated for competitive purposes. Hence, a notable growth in asset specificity (both tangible and intangible) can be traced to rapid economic growth and the increased value and velocity of assembler-supplier transactions. These have propelled Toshiba and other major firms into *network forms of organization* as a means to sustain high levels of interfirm learning and competitive performance.[5]

Invisible Assets and Network Organization

Take a good, yet fairly typical year, 1988, when nearly four-fifths of the production value of the Yanagicho Works was sourced from suppliers. Two hundred thirty-two of these suppliers sold $42 million of parts, components, subassemblies, and services to the Yanagicho Works, constituting about two-thirds of the value of the factory's output. Another 500 suppliers, give or take a few, sold an additional $6.9 million to the factory.[6] The value of supplier goods from all 732 suppliers, as a percentage of total manufacturing output, was 78 percent in 1988.

In that year, the high-performing, high-quality, cost-competitive products sourced from suppliers ranged from photocopiers to automatic mail-sorting devices and ATMS. Product life-cycles were short, often impossibly so—as little as six months in a few instances. Profit margins were slim, between 1 and 4 percent. Information and technology sharing between assemblers and suppliers was intense, sometimes resulting in little or no discernable differences with respect to product design, development, and manufacturing capabilities.

About half of the 232 key or core suppliers belong to Yanagicho's Kyoryokukai, or Supplier Association. Of the 232 suppliers, 164 provided *gaichuhin*, or ordered goods, whereas the remaining 68 made *konyuhin*, or purchased goods, following Banri Asanuma's (1985) classification scheme.[7] There were also *kanren kigyo* suppliers, or firms that are part of the Toshiba interfirm network at the corporate level of organization (as opposed to the factory level of affiliation).

Ordered, or *gaichuhin* goods, are custom ordered; that is, suppliers provide finished components and products against the functional specifications of assemblers. Extreme examples of customization are so-called "black box" parts, a situation in which assemblers may not know how suppliers met assemblers' functional specifications. In other words, design, development, and engineering are a "black box." In today's motor vehicle industry in Japan, black box parts run two to one as compared with others.[8] Suppliers' proprietary parts or assemblers' detail-controlled parts are the "others."

Suppliers vary according to the intensity and quality of the information they exchange with final assemblers. Core, or *gaichuhin* suppliers, as opposed to standard goods, or *konyuhin* suppliers, provide parts and deliver services according to design parameters set by final assemblers, but how they meet those requirements is not specified.[9] Suppliers of either sort are not directly involved in making the decisions that affect an assembler's strategy of product diversification and market penetration. But suppliers are an integral part of that strategy, because without them assemblers could not offer a full line of products in many different market segments, lower the risks associated with this strategy, and execute the strategy with speed, flexibility, and focus.

Hence, the governance system uniting Yanagicho and its suppliers is the key to Toshiba's competitiveness. Because Yanagicho does not lend capital to or own shares in suppliers, the Supplier Association is a organizational means for Yanagicho to ensure that its relations with suppliers are going well. "Going well" means that Yanagicho's intangible investments in suppliers are secure and well placed and that its 73–83 percent dependence on suppliers for parts, components, assemblies, and services is well considered.

The Factory and Managed Competition

Thirteen different product lines are made and managed at Yanagicho today. During the 60 years since its founding in 1936, Yanagicho has developed and manufactured scores upon scores of different products.[10] As distinct from a "mass production" factory, Yanagicho finds its mission in product and process innovation: in marrying new process technologies to existing product lines and in designing and developing new products with existing technologies. To some degree the variety of product lines at Yanagicho is a general feature of the Japanese electronics industry. Large firms produce a full line of fairly diverse products. In 1991, for example, Toshiba made 11,500 different kinds of products.[11]

Photocopiers and nonimpact printers are by far the most important products at Yanagicho, accounting for roughly half of the value of the facility's entire output. The labor-savings equipment department, itself a rather broad area encompassing ATMs and ticket-taking and cash-handling equipment, brings in about 27 percent. Optical-electric image systems, utility meters, refrigerator compressors, and specialty peripherals and parts account for increasingly smaller portions of the recipe.

Given Yanagicho's diverse range of products, 700 to 800 suppliers, and high numbers and value of outsourced parts, components, and subsystems, good supplier relations are a lot more difficult to achieve in fact than in theory. Developing, maintaining, and enhancing interorganizational capabilities means pursuing a dynamic balance between cooperation and competition with suppliers across a swath of product and process capabilities. The most important relations—those with core suppliers—become strategic alliances where partnering finds expression in interorganizational learning that strengthens the capabilities of assemblers and suppliers alike.

Suppliers are an integral part of Toshiba's strategy to offer a full line of products in many different market segments. Assemblers intensify their foci on particular product and market niches by recruiting other enterprises, usually smaller and less-well-endowed ones, to complement and complete their efforts. To the degree that market niches are exploited broadly and deeply, much of the credit must be assigned to the creative management of suppliers, that is, to the Supplier Association or Kyoryokukai.

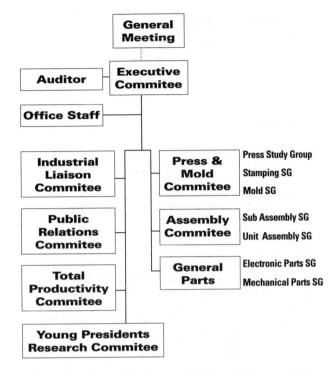

Figure 4.1   Organizing Yanagicho's supplier association.

As a formal means of interfacing with suppliers, the Supplier Association is a *site- and relation-specific governance system* designed to enhance the interactive capabilities of suppliers with each other and the factory. It does so by setting criteria for membership and by managing the incentives and rewards that underpin membership. The Supplier Association is a kind of network organization, involving a competitive-cooperative dynamic to sustain high-value-added contributions from suppliers and to maintain a basis for collective action with each other and Yanagicho.

Yanagicho's Supplier Association

The Supplier Association, or Kyoryokukai, was established in its present form in 1982. It existed previously as the Manufacturing Association (Seisanbukai), although the Seisanbukai was less active in its meetings and interactions than the present association, and the Seisanbukai did not stand on as equal footing with the factory as today's Kyoryokukai. There are 164 gaichuhin firms in the Kyoryokukai, or 71 percent of all core suppliers in the association. Although participation in the Kyoryokukai is voluntary, most firms join at the

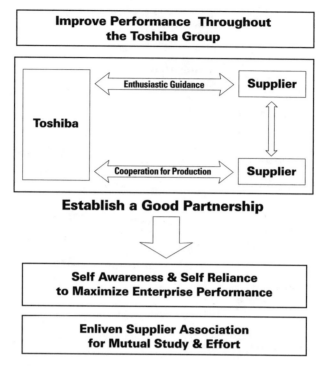

Figure 4.2   Philosophies of Yanagicho's supplier association.

invitation of the association and most of the invitations are made at the suggestion of Yanagicho's Purchasing Department.[12]

The association boasts a formal organization and a philosophy of partnership. While the factory is represented in the general assembly of suppliers, it does not participate directly in the functional and topical subgroupings of the association. However, the factory does participate in setting the agenda for monthly meetings, and through its socialization and training programs it has a considerable degree of influence on association activities. The formal structure and philosophy of partnership between Yanagicho and its suppliers are represented in Figures 4.1 and 4.2.

Given the incredible complexity of designing, developing, and manufacturing 13 different product lines within one factory, the 164 *gaichuhin* suppliers in the Kyoryokukai are certainly a major reason for Yanagicho's success. But dependence is not a one-way street: just as Yanagicho depends on its suppliers, the suppliers depend on Yanagicho for employment, sales, technological guidance, managerial aid, and logistical support. Interdependence is realized in spite of the large numbers involved. In the aggregate, at the corporate level, 1,300 firms and 59,000 people populate Toshiba's supplier network, figures that do not include the persons engaged in transporting, shipping, and distributing Toshiba products.[13]

It is necessary to distinguish among different types of organizational coordination to understand the processes of dynamic balancing that are at issue here. Yanagicho's suppliers are neither what are currently called horizontal business groups, *kigyo shudan*, nor their vertical analogue, *keiretsu*. *Keiretsu* are constellations of firms organized around a core firm or several core firms that own sufficient numbers of shares in them so as to induce a cooperative attitude. Financial ties may be further bolstered by ties of interlocking directorates and personnel exchange, and by history and traditions of common business practice.[14] However, the qualities that characterize business groups united by either horizontal or vertical ties, *kigyo shudan* and *keiretsu*, are quite different from the competitive-cooperative dynamic that defines supplier relations, especially those networks of suppliers mostly concerned with the sharing of intangible assets.

In truth, the amount of sharing and the degree of partnering that describe network organizations may vary considerably. The contemporary electronics industry presents a rather strong example of abundant sharing and effective partnering; witness Yanagicho's dependence on suppliers for 73–83 percent of the value of its output. High-value-added activities of this sort require close coordination, and in the absence of financial and contractual ties to secure and stabilize such activities, the Supplier Association, as well as resident engineer exchanges, TP campaigning, and after-hours socializing provide a basis for achieving high levels of coordination. There is considerable truth to the axiom of coexistence and coprosperity or risk sharing and profit sharing (*kyoson-kyoei*), a philosophy that Yanagicho espouses with its core suppliers.

However, coexistence and coprosperity are not without their context. What Toshiba does not need it will not buy, and therein comes the risk for suppliers—a risk not fully shared by big firms like Toshiba. Currently Toshiba is concerned that too much outsourcing may cause *kudouka*, or underemployment and unemployment. Even more worrisome, if Yanagicho depends too heavily on suppliers for particular products and services, Yanagicho may lose its creative edge in product-development innovation in those areas. That could result in reduced sales and lowered employment for assemblers and suppliers. In short, even if Yanagicho manages its suppliers, Yanagicho cannot be profitable without them. So, while Yanagicho decides the balance of cooperation/competition with suppliers, it is also evident that Yanagicho needs them as much as they need Yanagicho.

Toshiba-Style Contracts

For negotiating that balance, there are no Western-style contracts that fix in advance the terms of performance, namely, price, quality, and delivery. Yanagicho sets schedules for the delivery of quantities of specific items at specific destinations on specific dates. In theory, the "prices" for such "contracts" are negotiated every six months but, in practice, they are renegotiated every month as target prices. So, Yanagicho's "contracts" are in fact much more like broad purchase orders than contracts.

Schedules are fixed monthly, detailing daily deliveries of predetermined amounts for the next month. Prices remain fixed regardless of volume fluctuations during the month. Rather than specify prices by volume of parts delivered, the rules by which prices may be altered between monthly negotiations are specified. These are mostly changes in suppliers' labor, materials, and energy costs.[15] Because designs, products, and models are changing more or less constantly, prices are being renegotiated more or less continuously. At a minimum, new target prices are negotiated every few months, depending on the part in question and the volume being purchased.

Western-style contracts place little or no premium on the process of lowering production costs and developing increasingly better relations with suppliers. Prices remain fixed for the duration of contracts, generally six months in North American practice; more often than not, the same price point is used for negotiating a second six-month period. Maximizing profits, not maximizing information exchange and learning, is the aim.

Japanese-style contracts have opposite effects. For the short run, between price negotiations, suppliers are allowed to keep whatever cost reductions they realize, whereas in the long run, both parties benefit from new target price negotiations and cost reductions, given norms of coexistence and coprosperity. The success of suppliers in lowering costs is one of the factors motivating good relations, in addition to the frequency of design and model changes. Ultimately, buyers get less expensive parts, and suppliers increase their transactions with assemblers.

Put baldly, in exchange for bearing some risk of unpredictable production-cost increases, buyers benefit from the efforts of suppliers to lower costs. One could argue with this formulation, based on the proportions with which economic rent (profits) are distributed. However, given the valuation problems with sharing intangible assets, it would be extremely difficult to prove the point one way or another. Given the low levels of exit from the Supplier Association, one has to assume that profit distributions are sufficient to sustain supplier membership.

Thus, in spite of the small numbers of buyers, their monopsonic powers, and the arbitrary determination of the timing and content of contracting, price, volume, and delivery negotiations are regular, long-term, and market-based. More or less constant negotiation and renegotiation underpin a philosophy of coexistence and coprosperity, and long-term continuity in contracting allows for short-term indeterminacy in terms. Whatever was not quite right during the most recent negotiations will be fixed during the next. Processes of cooperation on the one hand and competition on the other imbue assembler-supplier relations at Yanagicho, in Toshiba, and more broadly in the Japanese electronics industry.

The Purchasing Department as Entrepreneur

The negotiation of purchase volumes, delivery terms, and prices paid to suppliers and the management of assembler-supplier relations are handled by purchasing departments (*shizaibu*) at Toshiba. The corporate purchasing department sets policies and procedures for the company as a whole through two corporate-level committees, the Manufacturing Strategy Committee and the Supplier Strategy Working Group. These bodies determine the nature and degree of (inter)dependence with suppliers; their work is probably more important to the long-term health of Toshiba than any other committees at the corporate level.

Throughout Toshiba, the balance between manufacturing in-house and using outside suppliers is strategically important. Toshiba is organized into numerous product groups, where divisions and business units are found. Each division and business unit has a Supplier Management Committee that sets policy directions and structures for supplier networks. Also, every manufacturing unit has its own purchasing department, charged with administering and overseeing factory and supplier interactions.

Suppliers have across-the-board lower manufacturing costs; labor, land, and overhead expenses are typically lower as well. Relying on suppliers for the manufacture and assembly of products means that the know-how associated with the design, production, and packaging of those goods may be lost to Toshiba. An ideal balance is somewhere between taking full advantage of the economies of specialization offered by smaller, more focused suppliers and not losing the knowledge accumulated by achieving those economies in-house. Finding that balance is an inherent part of the Knowledge Works model.

There are transaction costs associated with managing supplier networks as well as potentially large costs associated with a loss or transfer of proprietary know-how. Unless supplier partnerships are

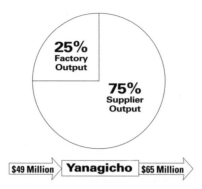

Figure 4.3   Yanagicho: Factory
and supplier value-added output.

effectively managed, mounting transaction costs can easily offset
the lower manufacturing costs that originally motivated the search
for outside suppliers. For example, if technically demanding parts
and processes are outsourced to less capable suppliers and considerable management and engineering oversight are required as a result,
transaction costs can balloon. Suppliers may also suffer higher rates
of personnel turnover than large firms do, so a lack of ability rather
than desire can drive up costs to unacceptably high levels.

Managing Partnerships

In 1986, the production value (internal transfer cost) of Yanagicho's
output reached approximately $65 million ($1 = 150 yen), of which
75 percent, or $49 million were goods purchased from core suppli-
ers. The average for all Toshiba factories in that year was 65 percent,
although the figure varies according to product line and factory
size.[16] The process and means by which Yanagicho manages relations with key suppliers are next discussed, using 1986 data. Given
the thoroughness and rigor with which this happens, the overall
importance attached to assembler-supplier relations can be easily
appreciated.

Of the roughly $42 million contracted by the Purchasing De-
partment, 4 percent went for basic materials, like carbon black and
hydraulic fluids; 33 percent was expended on *konyuhin*, or purchased goods, such as integrated circuits and other standardized
components; and 63 percent was spent on *gaichuhin*, or customized
products. Software development expenses, earmarked either for
products or production processes, are not included in these totals,
because they are embedded within administrative overhead for the
factory as a whole. It should be noted that both on- and off-site soft-

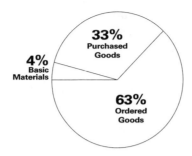

Figure 4.4   Goods purchased from suppliers.

ware costs are among the most rapidly rising for the factory. The breakdowns for purchased and ordered goods can be seen in Figures 4.4 and 4.5. Whole unit assemblies and subassemblies come from suppliers in some cases, like the six subassemblies for putting together ATMs; in these cases, the efforts of the factory are directed toward integration of subassemblies, testing and quality control, a final blending of software and hardware. As one department head told me, "Suppliers are really an extension of our own production lines."

"The Age of Selection"

Supplier management occurs within what is called the "Age of Selection" (*senbetsu no jidai*). A system to manage suppliers at the factory level of organization first appeared during the 1960s as orders outstripped the capacity of factories and, in response, factories cultivated suppliers for their extra productive capacity. Sourcing parts and assemblies from suppliers enabled firms/factories to reduce in-house tooling, engineering, and design costs at a time when investment funds for these were severely constrained. The first period of locating, training, and cultivating suppliers was known as the age of "nurturing excellent subcontractors" at Toshiba.

Since the late 1970s, however, an intensification of competition at home and a movement of production to overseas sites have led to less effort invested in developing good suppliers. Instead, techniques for distinguishing the good from the average are administered early on, and there is little tolerance for those who cannot make the grade. Competition is increasingly geared toward producing high-value-added goods. Under the leadership of President Futao Iwashita, who assumed leadership of Toshiba during the height of the oil crises in 1976, *sentaku keiei*, or "selective management," became a watchword.[17]

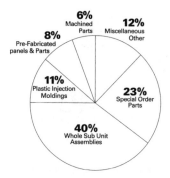

Figure 4.5 Breakdown of ordered goods.

The new "Age of Selection" policies practiced by Yanagicho's Purchasing Department was really an outgrowth of President Iwashita's "selective management" policy and a more recently initiated Total Productivity Campaign. The Purchasing Department believes that the number of suppliers will necessarily decline in today's tougher times, even while the relative value of supplier activities will likely increase. In fact, the number of suppliers has grown slightly in recent years. But this trend appears to reflect an increase in the number of product lines manufactured by Yanagicho rather than a heightened division of labor among existing suppliers. Since 1983, the number of core, or *gaichuhin,* suppliers has grown:

1983	146
1984	153
1985	160
1986	162
1987	163

However, in 1988, the Purchasing Department trimmed the number of *gaichuhin* suppliers in the Kyoryokukai to 151 in an effort to ensure their quality. Subsequently, the numbers have been further cut to 143 in 1989 and 131 in 1990. Although *gaichuhin* suppliers in the association have been weeded, the total number of suppliers of both *gaichuhin* and *konyuhin* has grown. Obviously, a tension exists between Yanagicho's desire to reduce the number of suppliers even as purchases of parts, components, and subassemblies grow. Given a need for extra production capacity and for a diversified range of technical knowledge and know-how, this tension is not likely to abate anytime soon.

Assembler-Supplier Relations

To survive in an "age of selection," both Yanagicho and its suppliers have made changes. For Yanagicho, this has meant a new personnel rotation policy, so that all departments move key persons through the purchasing function, thereby acquainting them with supplier management methods and problems. But even before the factory can effectively manage suppliers, there is a complicated process for deciding which parts, components, subassemblies, and assemblies to make in house and which to outsource. This is not as simple as it may sound, because Yanagicho believes that there are six advantages to outsourcing:

1. Suppliers can respond more quickly to market and technological change.
2. Alliances with suppliers promote industry progress.
3. Suppliers have lower wage costs.
4. Suppliers have fewer labor restrictions.
5. In the case of certain manufacturing processes, suppliers have special technological strengths.
6. Suppliers may have unused manufacturing capacity.

Obviously, managing suppliers is more than a simple matter of identifying wage rate and manufacturing cost differentials. Indeed, every part, product, and process at Yanagicho is scrutinized according to a complex formula that situates parts, products, and processes somewhere along a continuum of high-to-low desirability of in-house manufacture. The parts and products that fall clearly on one end of the continuum or another are not a problem; those that fall in the middle are.

Second order make-or-buy decisions have tactical as well as strategic elements. At the first level of decision making, clear-cut cost or technological advantages often lie with the factory or its suppliers. But a majority of cases are not so clear cut. Table 4.1 lists some issues in clarifying second order make-or-buy decisions. In addition to elaborate estimates of cost differences based on capacity-utilization calculations, additional considerations include estimates of wage costs, overtime, and fringe benefits for personnel, operating costs, repair costs, setup costs, depreciation costs, and insurance costs, as well as finance costs for materials, plant, and equipment.

Because a decision of what to make or to buy is itself part of an overall strategy of how, when, and where to attack the marketplace, Yanagicho's Purchasing Department must keep in mind other basic considerations. Although it may be possible to quantify many of these decision points, many of the most crucial issues are highly subjective and depend heavily on a feel for products and markets.

Table 4.1 To Make or To Buy

Decision Category	Check Point	In-House	Supplier
Quality			
Who has higher quality?	- defect rate -plant facilities -technical level		
Who has better quality monitoring management?	-level of routine quality control -level of work standards -inspection audit documentation		
Who has more experience and capability?	-years of experience -number of skilled workers (*jukurenko*)		
Delivery			
Who is more dependable?	-record on deliveries -efforts to improve deliveries -facilities location		
Who has a more cooperative attitude?	-efforts to improve delivery schedule -efforts to improve design specs		
Who has a better arrangement for managing deliveries?	-delivery management capabilities -following-deliveries system & efforts -balance of delivery & manufacturing abilities		
Other Considerations			
Amount of excess manufacturing capacity?			
Number of underutilized workers?			
Need for specialized training of workforce?			
Any increase in indirect staff?			

Yanagicho wants to keep its personnel, plant, and equipment utilized to the fullest extent, and thus to keep alive those products and manufacturing processes that have long production runs, will be profitable, and have high-value-added content. This often comes down to identifying the so-called key components of products, those parts and components that differentiate Toshiba products from all others. At Yanagicho, for example, these are the selenium drums for photocopiers, the casings for ATM machines, and the flat batteries for IC cards. Yanagicho is reluctant to let anyone else make these and, as a consequence, criteria for defining key components must be applied in all instances of custom outsourcing. The value of key components is such that Yanagicho makes them in-house, even at comparative cost disadvantage.

Generally, a supplier's manufacturing costs (based on their existing equipment and facilities) are remarkably lower, even with a healthy markup, than Yanagicho's. For this reason alone, everything possible is outsourced as long as Yanagicho's personnel are fully employed and production is running full and steady. Once parts and assemblies to be outsourced are determined, Yanagicho seeks at least two suppliers for every part and component. This fundamental rule of supplier management, a two-vendor policy, protects the factory from a variety of risks, such as an interruption in the flow of needed parts and components and a ballooning of outsourcing costs. Having two suppliers for the same part greatly minimizes these risks, and it also serves as a standard for comparing the quality, cost, delivery, and managerial competence of suppliers. Factories may also produce in-house what is outsourced as an additional check on quality, cost, design, and performance.

Comparative statistics are gathered by the Purchasing Department. The factory employed a six-part, one-hundred-point scheme for ranking suppliers in the past. Today, a two hundred-point scheme is used, weighing equally qualitative and quantitative factors. The schemes are reproduced in Figure 4.6.

This information can be graphed according to a conventional scatter diagram for illustration, with quality as the x-axis, for example, and technology or quality level along the y-axis. The aim of such illustrations, of course, is to increase supplier reliability, quality, delivery, and technical proficiency. Yanagicho expects better suppliers to follow the factory's lead in improving performance, lowering costs, improving quality, and generally advancing the design and manufacture of products. Indicative of the tighter fit that the factory seeks nowadays is the 30 points out of 100 that are awarded for supplier performance in delivery.

Comparative figures on supplier performance are gathered for every major product group. Although some suppliers may provide parts and components for several different product lines, most are specialized in just one or a few parts, components, or subassemblies. Depending on the nature of what is supplied, the level of acceptable performance with regard to quality, cost, or delivery can vary widely. The timing of deliveries, for example, can vary from 1 hour to 72

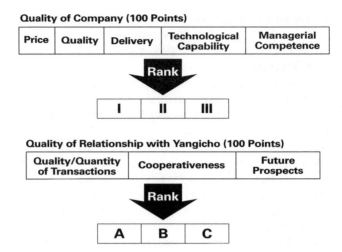

Figure 4.6   Supplier evaluation ranking schemes.

Item	Content	Evaluation							
1. Quality	Reject rate & attitude toward quality control	Rank:	1	2	3	4	5	6	7
		Points:	20	18	16	12	8	4	2
2. Delivery		Rank:	1	2	3	4	5	6	7
		Points:	30	25	20	15	10	5	0
3. Technological Level		Rank:	1	2	3	4	5	6	7
		Points:	10	8	6	4	2	1	0
4. Value Analysis	VA cooperation; VA suggestions: quality & number	Rank:	1	2	3	4	5	6	7
		Points	10	8	6	4	2	1	0
5. Evidence of Cost Down	CD drop/quarter CD drop/product	Rank:	1	2	3	4	5	6	7
		Points:	20	16	12	8	4	2	0
6. Cooperativeness		Rank:	1	2	3	4	5	6	7
		Points:	10	8	6	4	2	1	0

hours at Yanagicho, depending on the rates of production for different products. Supplier schedules are generally fixed three months in advance although minor adjustments are made right up to the current month.

Part of the cooperative-competitive balance is a tacit understanding that the factory will neither attempt to hire away the supplier's best people nor steal a supplier's production and delivery innovations. A large share of supplier motivations to lower costs and improve quality and delivery is the knowledge that such improve-

Table 4.2

Group I	Superlative suppliers that require little or no oversight; the group on which the future of the factory rests.
Group II	Suppliers who are excellent in their respective fields, even if they require some direction and advice.
Group III	Potentially good suppliers but needing considerable training and help at present.
Group IV	Suppliers who will lose orders during the next model change or mid-term negotiations, unless they make more effort.
Group V	Suppliers with whom transactions should be regularly reduced.

ments will increase their profit margins at the expense of contractors, at least until the next major renegotiation of price and delivery terms. Thus, suppliers work hard to lower costs and improve margins within the life-cycle of outsourced models and products.

The effort expended on appraising suppliers is not wasted. Using the standards for ranking suppliers discussed previously, the Purchasing Department has designed a method of grouping suppliers into five categories according to the degree of management effort that they require. The standards are listed in Table 4.2:

Comprehensive pricing and performance appraisals occur every three years on a rolling average. However, models and products are changing constantly, and so informal pricing and performance measures are almost weekly affairs. All suppliers are informed officially of their performance rankings once a year, but in fact, weekly and sometimes daily reports on the quality, cost, and reliability of parts, components, and subassemblies are tallied. The results are shared informally with suppliers as needed. Lists of "good" and "bad" suppliers, that is, those that are doing well or poorly according to factory's performance criteria, are posted publicly. Performance information is also given at monthly Supplier Association meetings.[18]

All Toshiba factories publish listings of their better suppliers, those falling into Categories I and II, and the information is circulated widely within the firm. When a factory wants to do business with another plant's established supplier, its Purchasing Department is expected to check first with the Purchasing Department having prior claim. Sharing information allows a review of the financial, technical, and managerial qualities of a potential supplier, thereby ensuring that a supplier is not overstepping its resources or that another factory will not be disappointed with a supplier's performance.

Suppliers falling somewhere in the middle grade require help in three areas: technical support, advice on how to lower costs, or information to improve general managerial skills. Yanagicho will gladly

Table 4.3 Ranking of Suppliers—The Yanagicho Works for 1986
(By Number of Firms)

Ranking	Number
Group I	20
Group II	16
Group III	17
Group IV	37
Group V	2

offer help in these areas, at no cost to suppliers, as long as suppliers appear willing and able to learn. Training of this sort can take place with the dispatch of personnel directly from Yanagicho as well as on-the-job training at Yanagicho. It can also occur through training sessions sponsored by the Supplier Association.

Supplier rankings are used by the factory to encourage, cajole, and criticize suppliers. At one point in 1986, the distribution of key *gaichuhin* suppliers (but excluding affiliated and *konyuhin* suppliers) according to the fivefold ranking scheme appeared as shown in Table 4.3.

The consequences of this ranking become immediately obvious when the value of the goods supplied to the factory is disaggregated by group classification. Not surprisingly, the higher the ranking, the more parts and components supplied to Yanagicho. Relative rankings and value of transactions can be seen in Table 4.4. Higher ranking suppliers get more work and make more money. Suppliers in the first three ranks number 53, or 57.6 percent, yet they produce 87 percent of the *gaichuhin* value of custom goods. Impressively, 20 suppliers in Group I provide the factory with nearly 60 percent of its custom-ordered parts and supplies. In short, more cooperative suppliers interact more profitably with Yanagicho.

Notwithstanding that some suppliers increase their business while

Table 4.4 Group Ranking and Value of Transactions for 92 Suppliers and the Yanagicho Works, 1986

Rank	Number	Percent	Value of Transaction (in 1,000,000 yen)	Percent
I	20	22	21,000.	58
II	16	17	1,800.	5
III	17	19	8,000.	24
IV	37	40	4,000.	12
V	2	2	200.	1
Totals	92	100%	35,000.	100%

others flounder, the link between Yanagicho and its core suppliers is not countercyclical; that is, less work is not contracted during a recession or an economic downturn. Core suppliers are expected to bear an equal share in any decline or rise in factory orders, and the Purchasing Department tries to make sure that everyone shoulders an equal burden. At least, this appears to have been true for the past 10 years.[19] (This is the informal assessment of Yanagicho's Purchasing Department.)

It is not for a lack of factory effort that some suppliers disappear. Yanagicho will send out technicians and specialists to improve supplier performance. Product and process information are freely given, and suppliers are welcomed to the factory for on-site training. But most of Yanagicho's suppliers are small, the majority with fewer than 100 employees, and with less than a million yen in assets.[20] Many are family-based firms. Generational changes in family leadership are common causes of at least some temporary difficulty. Some suppliers are stuck in a technology or they may be reluctant to make meaningful organizational changes, and the factory gradually disengages from these suppliers. Also, software as opposed to hardware contributes an increasingly higher value-added share today, and software development is done in-house as much as possible rather than outsourced. Depending on the amount of software associated with different products, therefore, some suppliers find themselves with an ever-shrinking slice of the pie.

Software products and services are in fact the most rapidly growing area of outsourcing (see Figure 4.7 below). The factory is not able to provide all of its key components and services internally in spite of a strategic desire to do so. On the other hand, growth in the number of suppliers and in the variety of goods and services that they provide are important sources of competitive renewal for Yanagicho. The factory's evaluation of suppliers needs to be considered in light of overall patterns of exit and entry by suppliers. Clearly the factory's efforts to manage competition among suppliers is matched by market forces that affect supply and demand for suppliers. These forces are beyond the pale of Yanagicho or the Supplier Association.

TP Transfer to Affiliates

On February 21, 1991, the Fifth Total Productivity (TP) Supplier Association (Kyoryokukai) Assembly of enterprises, working with Yanagicho in implementing TP practices, met in the third floor training hall of the engineering support services building at Yanagicho from 9:00 to 10:45 A.M.[21] Included in the two hundred or so persons in attendance were Yanagicho personnel and employees

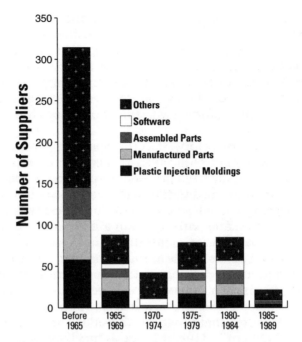

Figure 4.7 Initiation of supplier relation by
period and outsourcing category.

from other Toshiba operating units, as well as representatives of businesses supplying Yanagicho—and myself. On this particular occasion, five member firms of Yanagicho's TP Supplier Association
made presentations.

The head of the Supplier Association kicked off the assembly
with a simple statement of purpose. He summarized the goals of the
assembly as "mutual learning, mutual aid, and mutual benefit
[ryoho o shiri, ryoho o tasuke atte, ryoho o bengi hakatte]." He was
followed by the head of the TP promotion effort within the Kyoryokukai Association, the head of Yanagicho's Purchasing Department, and finally the five presentations by supplier firms. Yanagicho's factory manager, absent on this occasion, was scheduled to
speak between talks by the head of the Kyoryokukai's TP promotion
effort and the head of the Purchasing Department. The morning
program was organized for the benefit of Yanagicho's management
and presented by the TP Supplier Association.

The head of Yanagicho's Purchasing Department made several
noteworthy points during his three-minute presentation. The purpose of TP activities, he averred, was not to Toshiba's one-sided benefit but rather for "everyone's well-being." This happened when
"everyone acted in concert [issho ni natte katsudo suru]." "Good

suppliers are a Toshiba asset [yoi torisaki wa Toshiba no zaisan desu]," he concluded.

He was followed by a representative of the first of five firms making TP presentations that morning. The first speaker pledged that his company's aim was "to reach the same level of TP attainment as Yanagicho." As interpreted later by the head of the factory Training Department, the information presented at the Fifth TP Supplier Assembly was less important than the acts of presentation and participation. By making public statements of solidarity and compliance with Yanagicho's TP campaign, the five firms were pledging their efforts to be more central and integral to Yanagicho's operations. Efficiency would be enhanced by joining forces with Yanagicho even while their identity and profitability were in no way impaired by closer alignment.

At an afternoon meeting of the TP Kyoryokukai Assembly on March 12, 1991, 150 persons gathered to hear Yanagicho's factory manager declare, "We'll make our workplace better by our efforts [jibuntachi no shokuba jibuntachi no chikara de yoku suru]." In other words, the responsibility for improving performance was in our hands— the employees of Yanagicho and affiliates aligned with Yanagicho in the TP campaign. The factory manager pledged to help, by (a) periodically inspecting the shop floor environments that were linked together in production activities, and by (b) making Yanagicho central to divisional and overseas business activities. In this way, Yanagicho was attempting to position itself both as a domestic and international hub of business activities. Supplier participation in TP campaigning was an occasion and means for doing so; emphases were always on what needed to be done rather than on what had been accomplished.[22]

Sourcing Strategies: Champion and Balanced Line Variations

Partnerships with suppliers embody enormous opportunities. Not only do they offer a range of choices with respect to how products are designed, developed, and made, but they also influence what products are developed and when they will be made. Given Yanagicho's 232 core suppliers (not to mention 500 to 600 noncore suppliers), the factory boasts a combinatorial potential for product/process innovation and rapid product development and manufacture that far outstrips its proprietary, on-site resources. As always, the trick is deciding what to do oneself and what to do in concert with others.

In chapter 2, two different models for allocating available resources and capabilities were presented: the Champion Line and Balanced Line models. In the Champion Line model a single-product line dominates available functional resources—in research, design, development, manufacturing, marketing, and management. In the Balanced Line model, however, functional resources are more evenly balanced across a number of product lines. In the process of becoming a Knowledge Works, mobilizing factorywide resources in support of a champion line appears critical, even though resources and capabilities may become more evenly spread later.

Integration of various functions within product departments dictates the speed, quality, and reliability of Yanagicho's production function. While across-the-board integration is an outstanding feature of Knowledge Works, they cannot contain everything needed to survive and succeed in today's competitive rough-and-tumble. Supplier resources and capabilities have become all-important as a result. Integration of functions clusters around three modalities: applied research and design; "proving" concepts, prototyping, and design engineering; process engineering and production.

Supplier relations fall into three different patterns depending on the degree to which suppliers are integrated into these modalities. In the first pattern, characteristic of rapidly evolving, high-tech products, applied research, design, and development activities consume inordinate amounts of factory resources and capabilities. The development of photocopiers and automatic mail sorting equipment at Yanagicho illustrate this pattern.[23] Because the factory's attention is focused on upstream design and development activities, the primary role of suppliers is to provide production capacity.

In the second pattern, resources and capabilities are concentrated less on new product development, even though a champion line, like photocopiers or laptop/palmtop computers, may dominate factory-based R & D activities. In this pattern, responsibility for production as well as for "proving" production concepts is transferred to suppliers. Shared responsibility with suppliers for applied research, design, development, and production activities may emerge in this pattern, especially when product life-cycles require frequent product redesigning. This cospecialization pattern represents what is currently happening with many products, as when the development and manufacture of key components and subassemblies for photocopiers and ATMs have been shifted to suppliers.

The third pattern moves almost all of the responsibility for development and production activities to suppliers. This is characteristic of low value-added, relatively mature products where Yanagicho's applied research, design, development, and systems integration capabilities are not at all important. Because of the off-loading of such responsibilities to suppliers, factory resources can be more evenly

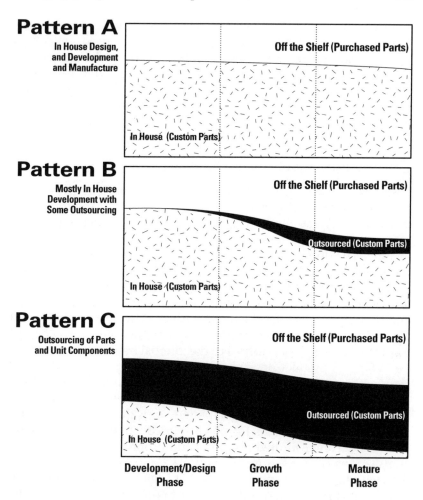

Pattern A

In House Design, and Development and Manufacture

Off the Shelf (Purchased Parts)

In House (Custom Parts)

Pattern B

Mostly In House Development with Some Outsourcing

Off the Shelf (Purchased Parts)

Outsourced (Custom Parts)

In House (Custom Parts)

Pattern C

Outsourcing of Parts and Unit Components

Off the Shelf (Purchased Parts)

Outsourced (Custom Parts)

In House (Custom Parts)

| Development/Design Phase | Growth Phase | Mature Phase |

Figure 4.8 Organizational patterns in the division of labor with suppliers.

balanced, as in the Balanced Line model. Yanagicho chooses this approach when product life-cycles are long and technical progress minimal, because suppliers invariably boast lower costs than the factory does. The utility meter line and some low-end, labor-saving devices, like automatic coffee dispensers, are good examples of this pattern.

Figure 4.8 captures the major distinctions between these patterns with respect to the division of labor with suppliers. All three patterns may exist simultaneously. Many different patterns coexist because there are many different kinds of products. As a result, Yanagicho's product department managers oversee an extensive division of labor without directly involving the attention and resources of the factory manager and his staff. Given a need for speed, reliability, and innovation in product design, development, and manufacture, points of balance between what is done inside Yanagicho and what is done outside are extremely important and always changing. A capacity to shift back and forth without losing competitiveness is absolutely strategic in the electronics/electrical equipment industry.

Integrating Suppliers Into Strategy

Toshiba's performance in three key areas are especially exemplary. Its governance system for supplier relations is one of these, and the other two are its systems for the internal transfer of technology and for new product development. The first of these is the subject of this chapter. The other two are described in chapters 3 (Organizational Campaigning) and 5 (Computer-on-a-Card).

Obviously, good relations with suppliers are interconnected with arrangements that are also important for the transfer of technology and effective product development. Toshiba's strengths in these areas are critical to its manufacturing strategy, and they are directly related to the Knowledge Works form. For the internal transfer of technology and effective product development processes, however, good relations with suppliers require Toshiba to span numerous organizational boundaries that separate it from its *gaichu* and *konyu* suppliers.

In order to attract leading suppliers, forward-looking, strategic investment on the part of Toshiba is needed. Suppliers do not want to squander their limited resources and proprietary know-how on transactions with just any company. The opportunity to be associated with big projects incorporating cutting-edge technology are inducements to attract the best suppliers. In order to do so, since 1984, Toshiba has invested a steady 15 percent of the value of its manufacturing output in plant and capital equipment investment.[24]

High levels of capital investment continued even during the economic downturn of 1986–91. The investment was not directed simply toward enhancing manufacturing volume and variety. For example, 250 billion yen went into state-of-the-art integrated circuit design facilities at Toshiba's ULSI (Ultra-Large Scale Integration) laboratory; 100 billion was invested in Toshiba's SSDC (Semiconductor Systems Design Center) for CAD (computer aided design) and Super Computer systems. These two investments strengthened the Semiconductor Group's R & D capabilities, directly impacting the development of hundreds of different products using Toshiba semiconductors, like Yanagicho's photocopiers and laser beam printers.

Toshiba poured investment into the Oita Works in Kyushu to produce leading MOS (metal oxide semiconductor) memory devices, both 256K and 1 megabyte devices. Design, development, and manufacturing capabilities were concentrated at Oita so as to maximize cross-functional and scale-based learning. The latest stepper and etching process equipment were introduced early in the manufacturing process. By concentrating and integrating investments in robust manufacturing sites, the Knowledge Works model, Toshiba became a leader in the production of 1 megabyte chips while a few years earlier, it had to be considered a follower in the manufacture of 256K chips.

Toshiba's catch-up strategy was to concentrate all short-term R & D (less than three to four years) at one location, the Oita Works, in order to gain the maximum benefits from siting design and development along with production. More advanced R & D, four to seven years, or two product generations away from production, on the other hand, is concentrated at Central R & D and ULSI Labs. In other words, when the technology under development is little more than one product generation away from application (that is, three to four years from full production), it is transferred downstream to SSDC (Semiconductor Systems Design Center) Labs or to Knowledge Works, like the Oita Factory.[25]

Toshiba's approach to new semiconductor development is rather different from Hitachi's, Mitsubishi Electric's, and even NEC's, Toshiba's domestic rivals. In the case of Mitsubishi Electric, near-term semiconductor design, engineering, and manufacturing occur in the same division, like Toshiba, but unlike Hitachi and NEC where different divisions are responsible for semiconductor devices earmarked for different size or purpose computers. So, in the cases of Hitachi and NEC, advanced chip design and development are separated organizationally from applied chip design and development.

In Toshiba's and Mitsubishi's cases, the top-down transfer of technology from advanced research and development is contained organizationally within the same division. In Mitsubishi Electric's case, however, there are fewer engineering labs paired with multifunction, multiproduct factories, and as a result, Mitsubishi Electric's factories are less robust with respect to the creation, integration, and diffusion of knowledge. Given that less is transferred down and created on-site, less is available for swapping, augmenting, and diffusing with suppliers. The success of the SuperSmart card project at Yanagicho provides a graphic illustration of the impact of these differences, as detailed in chapter 5.

Managing Competition and Network Governance

Japanese industrials are only as good as their suppliers or so this chapter argues. Today, good relations with key suppliers have become absolutely essential to manufacturing success. Given the importance of suppliers, Toshiba has organized Supplier Associations, TP training for suppliers, and a host of other initiatives to supplement, and on occasion, supplant market-based means for managing suppliers. The importance of such governance arrangements is related to the degree of asset specificity shared by Toshiba and its suppliers.

The greater and more important the sharing, the more attention is given to managing those relations. This is especially true when intangible knowledge and know-how are shared, especially with core suppliers. With them, supplier relations are highly sequenced, structured, and strategically driven, and the mechanisms of coordination are neither the "arms-length" haggling of the marketplace nor the administrative fiats of Toshiba—neither markets nor hierarchies, as Woody Powell (1990) declares.[26]

Instead, network forms of organization join otherwise independent suppliers in voluntary, mutually beneficial, cooperative activities. This characterization describes supplier relations at Yanagicho, especially the relations between the factory and its best core (*gaichu*) suppliers. Voluntary, mutually beneficial, and cooperative relations are realized within an organizational context that allows for bargaining, voice, and exit. Organizational boundaries are neither entirely open nor completely closed; they are "permeable."[27]

Permeable boundaries, a large number of actors, a mix (but not an amalgamation) of marketlike and hierarchylike features, and managed entry and exit of suppliers culminate in a design that can only be described as a network organization. Toshiba has no financial stake in any of Yanagicho's suppliers, except for a few corporate affiliates (*kanren kigyo*) that are coincidentally factory suppliers. Moreover, only 7 percent of 232 core suppliers are dependent on the factory for more than 70 percent of their business, and only 10 percent, or 23 firms, for more than 50 percent of their business.[28]

Low levels of dependency are not accidental. A decided dependence would be too risky in rapidly changing market circumstances. So, neither financial nor contractual safeguards appear to regulate transactional flows across the permeable boundaries of Yanagicho's supplier network. If not investments and contracts, what then? Governance with an eye toward an effective management of "mutually held" assets has evolved instead. The specific assets in question are knowledge, know-how, and the embodied capabilities of a network organization of assemblers and suppliers.

Co-ownership and co-management require a form of organization different from either market-based, price-bargaining arrangements or vertically/quasi-vertically integrated firms. Neither is well suited to the long-term partnering, highly complex dance routines, and perishable asset-specificity that motivate assemblers and suppliers in the fast-moving electronics industry. Instead, network forms of organization promote a dynamic balance of cooperation and competition, interdependency and innovation that can be observed narrowly at Yanagicho and more broadly throughout Japan. The resulting combination of highly focused factories (and firms) and networks of suppliers fosters a dynamic tension in assembler-supplier partnerships, or what might be called "managed competition".[29]

Strategies of managed competition economize on transaction costs (partnering transforms some portion of transaction costs into learning or experience costs) yet maximize the independence and autonomy of individual actors (suppliers and assemblers may choose the degree and amount of cospecialization, a range-of-choices model). By economizing on transaction costs and by amplifying learning through sharing, network organizations encourage timely and efficient use of resources. Herein lies the importance of supplier grading schemes, training seminars, resident engineering programs, and all the other activities and methods that animate Supplier Associations. These are *roughly proportional* to the degree of cospecialized assets, and hence, they are tied to the prevalence and performance of network forms of knowledge-sharing.

High levels of cospecialization necessitate particularly intense, frequent, multilateral information exchanges that, in turn, constitute a basis for collective action. For example, when Yanagicho developed its line of automatic mail sorting equipment in the 1970s, it made an early decision that suppliers would handle the CNC-based metal punching and wire cutting subassembly operations. Thereafter, Yanagicho could not produce mail sorting equipment without the support of suppliers in these areas, and suppliers would not invest more in CNC equipment without increased outsourcing by Yanagicho.

When asset cospecialization is less well developed, less robust coupling and governing mechanisms will suffice. Interlocking directorates, executive exchange protocols, and periodic bank audits of debtor performance are examples.[30] In Yanagicho's case, however, suppliers cannot provide what the factory wants without its active collaboration, detailed orders, design advice, technical and managerial assistance. Even more important, the factory's Purchasing Department seeks only the best efforts of the best suppliers. Those who do not measure up are edged out.

Attitude, action, and performance are all carefully assessed. Of the 300 enterprises that supply Yanagicho with most of what it buys, the top 50 get 80 percent of the business and the top 90, close to 90 percent.[31] Top suppliers are obviously required to make transaction-specific investments to meet and maintain supply conditions. The best suppliers are partners in a division of labor that shares risks and rewards. That worthy symmetry is earned by suppliers' strict attention to manufacturing detail and costs as well as a willingness to compete against other suppliers. Core, or *gaichuhin*, suppliers experience close and constant supervision, while "purchased goods" are simply bought according to the lowest bid, although there may be some effort to move more sophisticated suppliers of purchased goods into the ordered goods category. Related factories and affiliates receive little attention. They have their own built-in systems to ensure high-quality, low-cost products, and they are expected to be good yet autonomous.

Managing Competition and Cooperation

Both competition and cooperation further the ends of assemblers and suppliers. There are some tasks that benefit from competition, like demanding quality and high productivity among suppliers; there are some that rely more on cooperation, like the joint sourcing of hard-to-get materials or a sharing of manufacturing knowledge. There may even be a hierarchy of such goals, and for the sake of uppermost goals, like the marshalling of organizational capabilities, lesser goals, like short-term profits, may be subsumed or sacrificed.[32]

Managed competition of this sort infuses and enriches supplier networks, yet it is manifest only in the context of a governance system that joins and mutually motivates assemblers and suppliers. That is the function of the visible hand of management, the Supplier Association. As a result, Yanagicho can choose interorganizational goals and aims in process rather than as fixed structures or policies. In the fast-moving electronics industry, such abilities make all the difference to competitive survival and performance.

In speed there is also safety for suppliers. If Yanagicho could do everything internally: design, develop, and make most or all of its needed parts, components, and subassemblies, then suppliers would be more akin to subcontractors—firms that merely sell what the factory chooses not to make. Partnering would be unimportant. But speed of competition means that it is organizationally difficult and financially foolish for Yanagicho to do everything. Sharing risks and rewards with suppliers makes good sense in a fast-moving, topsy-turvy world.

In competitive circumstances where Toshiba buys 90 percent of the value of its photocopiers from suppliers, the ability to outsource parts, components, and subassemblies quickly and reliably constitutes *a relation-specific organizational capability*.[33] That is, Yanagicho's site-specific capabilities in photocopier design and manufacture are tightly coupled with its relation-specific network of suppliers. Supplier networks of this sort do not represent some feudalistic remnant or cultural peculiarity of the Japanese. Instead, supplier networks arose as a way to respond quickly, effectively, and well to changing market and technical trends. Large firms did so by intensifying their foci on particular product and market niches and, at the same time, by recruiting other enterprises, usually smaller and less well capitalized, to deepen and broaden these foci by supplying specialized know-how, process skills, and production capacity.

Managing "inside" and "outside" capabilities is the competitive need today. Attitude and learning are the keys to doing so. For suppliers, these lead to greater position and more power within the Supplier Association; for Yanagicho, they represent the means to deliver the right products, in the right amounts, at the right prices, to market on time.

By nurturing a factory-centered supplier network of some 700 to 800 suppliers, Yanagicho simultaneously solves two problems: how to get the *best* prices for products and services, and how to get the *right* products and services. Both are available in a network of suppliers that is at once responsive to factory needs and, at the same time, replete with its own capabilities. Though not an economist's ideal market of many buyers and sellers, this arrangement offers information sharing (responsiveness) as well as choice (multiple sourcing). Because the system mimics the market, both right prices and right products are part of the picture, which explains Yanagicho's success in such diverse markets as photocopiers, laser beam printers, and railroad/turnpike fare systems. Today, more than ever, Yanagicho and Toshiba are only as good as their suppliers . . . and Supplier Associations.

FIVE

Computer-on-a-Card

Between December 1985 and September 1987, a rather brief period of 22 months, the Yanagicho Works designed, developed, and delivered an engineering sample of the world's first SuperSmart card: a full-function, on-line/off-line, integrated circuit (IC), 16K RAM, three-line display, 10-key, credit-card sized computer.[1] The speed with which this pathbreaking product was fashioned, given the formidable design, prototyping, and manufacturing hurdles, bespeaks a means of product design, development, and delivery that is unusually effective. This chapter highlights Yanagicho as a product development platform, underscoring its versatile product and process capabilities and its network of affiliates and suppliers. Site- and relation-specific assets like these are integral parts of the SuperSmart card and Knowledge Works stories.

Smart Cards

We live in an age of credit and debit cards, and this story begins with that reality. Visa International wanted a credit card to end all cards: a card with enough internal memory to keep a running balance of multiple accounts; a card operated with a secret PIN number, thereby reducing fraudulent and criminal use; a card that could be used on-line or off-line according to user needs; a card with self-authorization features, that is, one that would allow transactions to be processed without an external reader/writer terminal. Visa International wanted a card unlike any other card in the world.

Visa coveted a multipurpose card combining credit, cash, banking, memory, and prepayment features because such a card would turn Visa International into a household name, giving it an even larger slice of the huge credit card market. With a total of 1.25 billion cards in circulation worldwide for all credit card companies, any significant increase in market share would be translated immediately into huge profits and international prestige.[2] Visa's wished-for card, when used with point-of-sale (POS) terminals or touchtone telephones, would provide even more features, like automatic dialing, reservation, and remote settlement of charges. The card would be an entire electronic funds management and transfer (EFT) system.

Revolutionary in its capabilities, Visa's card would have the same physical dimensions as a standard, plastic, magnetic stripe card. Given its unprecedented features and demanding specifications, and knowing that firms capable of making such a card would not be lining up outside its San Mateo, California, offices, Visa International pitched the project to a number of leading industrials worldwide. But as Visa discovered, in 1985 no one anywhere had the discrete technologies and systems integration capabilities needed to make such a card, though a number of French firms with government research grants had initiated preliminary development work on IC banking and telephone cards. The original patent holders, Michel Ugon and Roland Moreno, worked for the Bull Group but, unfortunately, work on IC bank cards in France was cancelled when the banking industry was nationalized by President Mitterand in 1981.[3] Nonetheless, IC telephone cards (as opposed to bank cards) became hit products in France, where they rescued the public telephone system from the pervasive petty thievery and vandalism that had plagued coin operated telephones since the early 1960s.

Visa approached 3M in the United States, Bull in France, and Toshiba in Japan. Visa wanted at least two vendors to minimize the risk of interrupted supply. Ideally, it sought separate sources of supply in each of the world's leading credit card markets, namely North America, Western Europe, Asia and Japan. But except for one possibility, Mr. Pino Franchini, general manager of Visa International, came up empty handed in his search for a maker of SuperSmart cards. Everyone but Toshiba refused. In fact, Visa had more than a passing interest in Toshiba because Toshiba held the "Visa" trademark as one of its many trademarks in Japan. Without rights to its trademark name in Japan, which Toshiba ultimately relinquished, Visa could not put its logo on thousands of airport, luxury hotel, and downtown ATMs.[4]

But the "Visa" trademark was the least of Visa and Toshiba's worries. The card's components—the LSI (large scale integrated chips), LCD (liquid crystal display), transducer, ultrathin battery, and waferlike packaging—were not available anywhere, except perhaps for Pentagon or other military users. No one was imprudent enough to claim that they had the systems engineering know-how to marry unavailable hardware with unheard of software. Actually, Toshiba had some experience with IC cards, as it had filed a patent in 1977 (which expired in 1992) for an offset-mounted, plastic, IC card. Since 1981 Toshiba had also been helping Mitsui Bank develop an IC banking card system that was first tested in October 1984, a first in Japan.

But these were relatively minor triumphs in the face of Visa's triple challenge: simultaneously developing hardware components, software systems, and the assembly process know-how for the world's first computer-on-a-card. In spite of these challenges, within Toshiba, Yanagicho was determined to steal away the project from the Ome Works, another factory producing laptop computers and microcomputer peripherals. Given Ome's computer design and manufacturing experience, an IC card project might have gone there more logically. And their marketing experience with laptops might have proven very useful in estimating market demand. But, no one seemed especially concerned with these issues in 1985.

While the technical challenges were certainly formidable, a so-called "smart card" offered obvious and numerous business advantages. Security was a huge attraction. Fraudulent card use was costing firms like Visa International and MasterCard over $300 million annually in 1990. Adding counterfeit losses and credit overcharges jumps the totals to $2 billion.[5] When such losses are multiplied by even a handful of major credit card issuers, like Visa, MasterCard, American Express, and Carte Blanche, among others, total annual losses are staggering. With smart cards, however, only persons with proper authorization or PIN numbers could incur charges. When hooked to relatively cheap off-line authorization terminals, PINs could be quickly checked, further reducing fraudulent use.[6] In short, a smart card's self-authorization features allowed it to be used at Visa's 45,000 ATMs but not be sold on the street for illegal use.

Beside credit losses, another factor driving Visa International was the deregulation of the telecommunications industry. At that time, the mid-1980s, telephone rates were expected to go up as much as go down after deregulation. Increasing rates would alter the economies of point-of-sale terminals, making the off-line capabilities of an IC card, its PIN and encryption technologies, cost-effective alternatives to rising, on-line management charges.

When used as an EFT card, smart cards do not even require on-line authorization. This saves on hardware and software installation and maintenance expenses, not to mention on-line telephone charges and the costs of authorization terminals. Because of the limited memory envisioned for first-generation cards, however, they would have to be "charged up" periodically. This could be accomplished via existing POS and banking facilities on a weekly or monthly basis, depending on the frequency of card use and credit limits. Planned functions for Visa's smart card are given in Figure 5.1. In the late 1980s, smart cards looked like a high-end solution to such low-life problems as theft, fraud, and other illegal and inappropriate uses.

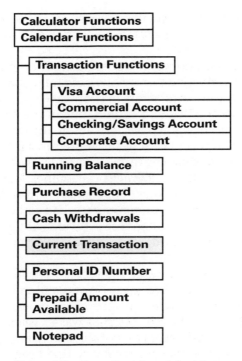

Figure 5.1 SuperSmart card functions.

They also offered Visa International and its card users a high-tech cachet, an image of money and power, that appeared unbeatable.

Product Champions

Two Toshiba managers, one in the Business Machine Division and another at Yanagicho, championed the development of the Toshiba SuperSmart card. The story unfolds with Mr. Mitsuo Kubo, head of the Automation Equipment Business Unit within the Business Machine Division, sending a memo on October 25, 1985, to Mr. Hiroyuki Matsuda, head of Yanagicho's Purchasing Department. The memo informed Mr. Matsuda that an ATM salesman's chance visit to Visa's headquarters in California had resulted in a potentially huge business opportunity. Visa International wanted a so-called smart card and they wondered if Toshiba could make it.[7]

But even as Toshiba managers were boasting to Visa International that they could make a smart card, only a few managers and engineers actually believed that they could do it. Yet, it was precisely the esprit de corps and can-do attitude of a few good managers and engineers that distinguished Toshiba from other firms that Visa International approached as development partners. Toshiba and Yanagicho had not lost the entrepreneurial spirit and risk-taking attitude of smaller firms, plus Toshiba had a trademark and some encryption software that Visa wanted.

Yanagicho's bravado was understandable if foolhardy. Yanagicho boasted a long and proud history of successful product development beginning with the manufacture of shortwave radio tubes and chassis before the Pacific War and culminating in the development and production of electronic typewriters, automatic mail sorting equipment, photocopiers, and laser beam printers in the 1960s and 70s.[8] Yanagicho combined precision machining know-how with microelectronic design and development capabilities. But in all fairness, nothing on the order of Visa's SuperSmart card had ever been attempted at Yanagicho: key components had to be developed and miniaturized; new card and on-line software written; assembly process technology developed; and systems integration engineering know-how created.

Kubo's memo to Matsuda suggested that the project might (should?) be based at the Yanagicho Works and that Matsuda might (should?) begin to assemble a development project team. Matsuda was a good choice for the job. Matsuda was 47 years old at the time. For four years, he had been head of the Purchasing Department at Yanagicho, a key function for a factory that bought two-thirds to three-fourths of the manufactured value of its products from suppliers. He was trained as a chemical engineer, had logged eight years as an IC chemist, eight years in calculator technology, two and a half years in corporate engineering (value analysis), and three years in Purchasing at the Fuji Works before coming to Yanagicho.[9]

For three months, from October 1985, until the official launch of the project in January 1986, Matsuda kept preparations mostly to himself while he readied a checklist of people and resources that could be called upon, once (and if) the project was approved. Matsuda's incipient groundwork was critical in the decision to give the project to Yanagicho rather than to the Ome Works, perhaps a more sensible choice as a smart card development site. Kubo argued forcefully on Yanagicho's behalf at corporate headquarters because Matsuda and Yanagicho were already masterminding a "flying start."

Flying Start

Initiating a project before it begins officially or launching with a flying start can be a risky business. Focus is divided because energy and effort are being diverted from other uses; politicking for the right plant, equipment, and people begins before there is official approval; designers and engineers are nervous over who's on and off the team.

Everyone is anxious to get going even though it is not yet certain what the project is, how big it is, when it will go, who will be involved, and how many resources will be committed. Indecision and uncertainty can derail or even kill a project before it ever gets going; when Toshiba nervously announced the venture in December 1985, it was not even holding a signed contract with Visa International. That came a month later.

The inability of Visa International and Toshiba to agree to contractual terms early on combined with a flying start proved a fatal combination. Toshiba was making a strategic commitment to Visa in advance of normal guarantees. And even if Toshiba could deliver the goods—a big gamble—would smart cards actually appeal to credit card users? That, of course, was Visa International's responsibility.

Flying starts normally bring many advantages. Since gaining consensus among many actors is time consuming, flying starts allow consensus building processes to begin early, permit preproject planning to smooth the way, and save on total elapsed time and costs by speeding up what might otherwise be a long and difficult development process. Front-end activities include a preselection of vendors, up-front commitment to vendor production resources, and as a result, a focus on in-house design engineering, testing, and tooling.

The complexity of the SuperSmart card design and the unpredictability of the development process precluded the mobilization of all the needed resources for the project by Yanagicho alone or, for that matter, even Toshiba alone. There was just too much to do, in too short a period, and too much of it was outside of Toshiba's proven core competencies. In other words, the project was extremely risky. Early discussions with vendors allow a head start that may pay real dividends in the speed and costs of development later on. Yet, in this case, as in most flying starts, outsiders are completely dependent on the know-how, experience, and assessment of those on the inside. The future of the SuperSmart card depended on Yanagicho's relation-specific capabilities with vendors.

The specifications called for a laser-welded, stainless steel sandwich of 0.76 mm (± 0.08) width, holding an LCD and crystal oscillator of some 0.5 microns. A surge-absorbing device was needed as well to protect the card from static electricity and from unauthorized software decoding. Stainless steel as opposed to plastic was necessary in order to protect the sensitive components housed within the card, although by the time the card went into production in 1988, a flexible stainless steel card was under development.[10]

Both the inside and outside halves of the printed circuit board (PCB) were used to mount circuitry and components. Double-sided PCB technology was just coming into vogue in the mid-1980s and

Toshiba had some experience with the technique. Hence, there were four layers to the card: two stainless steel and two PCB layers. Sufficiently thin LCD, IC, transducer, and battery were obvious development bottlenecks because they were generally unavailable, anywhere in the world, at the time. The bonding strength between the IC chips and the PCB was another area of difficulty, given that a single strand of 23-micron gold fiber was supposed to connect three chips and one diode to the board.

These were substantial bottlenecks, requiring real innovations in product, process, and materials design and development. By way of comparison the degree of innovation—really improvement—required during the development of the cash handling and ATM line, other products designed, developed, and manufactured at Yanagicho, was minimal. But incremental improvement would not make the grade for Visa's SuperSmart card. In 1985, neither Toshiba nor Yanagicho boasted the complex and precise capabilities required for the project.

Inside/Outside Project Positioning

The technical and systems engineering complexity of the card required Yanagicho to forge alliances with other Toshiba units and outside vendors. Toshiba moved swiftly to secure these by initiating discussions with vendors for key components, such as the ultrathin battery, three-line LCD, and double-sided PCB. Such discussions are not as simple as they might sound. Outside expertise in critical design and development areas are crucial to project success yet Toshiba could ill afford to lose control at the outset. For success to be guaranteed, the key components should be Toshiba's alone to supply. A way to reduce risks in this situation is to contract with outside suppliers while continuing with internal development activities. Then, if one side fails to deliver, the other still can.

However, several key components for the SuperSmart card project could not be made by Toshiba, such as the ultrathin film battery, the crystal oscillator, and the two-sided PCB. The first of these, the thin film battery, was fortunately made by a Toshiba affiliate, Toshiba Battery, and so sourcing was not a major problem. But in 1987 a crystal oscillator and two-sided PCB technology appropriate for a prototype card were available only from the Seiko Corporation. Seiko was neither a member of the Toshiba group of companies, a more or less vertically integrated industrial group, nor related to the Mitsui Bank business group, the larger horizontal group of firms to which Toshiba is connected.

Despite the absence of an established relationship with Seiko,

Toshiba still reckoned that it was less expensive to buy rather than attempt in-house development of crystal oscillators and double-sided PCBs. Also the time required to develop such key components, not just the risk, prompted Toshiba to seek an accomodation with Seiko.[11] The same logic held in terms of Toshiba's licensing agreement with Data Card Company, headquartered in Minneapolis. Data Card enjoys the largest market share for card embossing machines in the world. By licensing their technology, Toshiba ensured worldwide compatibility for the card's embossing system, another risk reduction move.

In any project, the quality of suppliers, their reliability, technical capability, and willingness to work closely with project managers, is far more important than prices, because during early development stages, full and complete information exchange between all parties drives progress and invention. Pricing is problematic in early development when it is impossible to estimate accurately how much time and money will be needed to design new circuitry and componentry. How do you price new knowledge for which a market value has not yet been determined? How do you protect intellectual property rights among friendly yet independent firms? Securing ownership rights for new and radically different components that are codeveloped with suppliers is rather different from contractual arrangements made with suppliers of conventional outsourced products.[12]

Some portion of development costs can be recovered if there are alternative markets for components. But if parts and components cannot be sold separately, development costs have to be fully recovered in the prices suppliers charge Toshiba. Final pricing can be decided later on, of course, after a certain amount of design and development work has been completed, as long as good relations are forged with suppliers early on. While flying starts may be helpful in this regard, reputation effects are the real clincher.

A reputation for fairness and technical virtuosity attract independent contractors, even though fairness in developing entirely new technologies, components, and their applications are hard to define and harder yet to affect. Hence, good political relations with suppliers must be buttressed with a strategy of paying special attention to intellectual property rights. Where conventional know-how originates with assemblers, past experience with suppliers may provide a means of cooperation even if the legal parameters for doing so are not clear. When new, key technologies originate with suppliers, however, assemblers are often prevented from using the know-how until everybody's legal position is clarified. The way for final assemblers to protect their legal position vis-à-vis suppliers is patent registration.

By registering its own patents early and often, on the other hand, Toshiba could release information to suppliers at the very start of the project, lengthen the time for conducting feasibility studies, and relax the political atmosphere within which information exchange and cooperation transpire. Overall, Toshiba filed 1,000 patents in the course of the SuperSmart card project. Robust patenting strengthened Toshiba in its negotiations with suppliers, knowing that technical bleedthrough to vendors would be mostly contained by legal codices and political goodwill.[13] And everyone loves a winner. The large number of patent registrations drew suppliers like honey draws bees.

All-for-One and One-for-All?

The SuperSmart card project was designated as an "across-all-divisions" project when a contract with Visa International was signed on January 16, 1986. "Across-all-divisions" means that at least half of the money for the project comes from the corporate head office as opposed to operating divisions. The designation is critical, especially in areas of new and risky technical development, because without it, major corporate resources cannot be committed to a project. Generally, development projects are financed by sectors and divisions, one or several levels below the corporate head office.

Even while much of the funding for the project came from the corporate headquarters, control and management of the project funds were located in the Labor Savings Automation Equipment Product Department at Yanagicho, itself a suborganization of the Automation Equipment Business Unit of the Business Machine Division. Additional funds were funneled to the project through the Information Technology Division, the Yanagicho Works Research Lab, the Yokohama Manufacturing Engineering Center, as well as from the IC Card Systems Department created at the Yanagicho Works. In other words, once the project was recognized as one of high strategic value, a separate budget line from many sources, corporate as well as operational, was created.

The "across-all-divisions" imprimatur was important for signaling that the company was behind Yanagicho's gamble. Without this, it might be difficult to mobilize resources and to minimize transaction costs between different factories, divisions, labs, and human resources, within and without the firm. Also, a high-level blessing enabled top-level executives to be more closely involved in monitoring the project. Senior and executive directors of the firm, like Shoichiro Saba and Yoshinobu Sato, would not normally be watching a single development project's progress in one of Toshiba's two dozen or so factories.

Masuo Tamada, a pattern recognition specialist with electrical engineering training, who took over as head of the IC Card Systems Department after Matsuda was promoted to factory manager, believes that three factors motivate technicians and engineers to join a project like the SuperSmart card project.[14] First, they have the opportunity to learn product development skills. In factories devoted to product and process innovation, participation in development projects is one of the surest ways of getting ahead. So, in Toshiba's way of working, getting ahead and risk taking are intimately tied to project assignments in Knowledge Works.

Second, given the localization of project resources in factory sites and the speed with which they are mobilized and integrated once the green light is given, one's future and face are tied irrevocably to project success. Participating in development projects becomes a variation on military style, up-or-out promotion systems. You become marked, known for delivering the goods or not. The consequences of either reputation are fairly predictable. Especially when it is likely that you will return to the department, factory, lab, or division from whence you came before a development project assignment.

In short, participation in development projects offers employees a combination of high rewards and high risks. The rewards are opportunities to learn product development skills, to enhance one's reputation, and to earn additional income by helping to bring new products to market. Project team members' salaries reflect a portion of the additional revenue generated by successfully developing new products.[15] On the other hand, a bust is a bust. If your performance does not measure up or if the project fails, a reputation for failure lives on. All in all, the combination of high rewards and high risks is a good one, according to Tamada in 1990. Engineers need to take risks in order to get ahead.

A risk-taking mentality is especially important for large firms like Toshiba that offer some version of lifetime employment and seniority-based compensation. With employment and remuneration conditions like these, the likelihood of diminishing returns among mid-career and senior employees looms large. By encouraging participation in development projects, especially in Knowledge Works, firms motivate the more ambitious and creative employees to learn more and build their reputations, and thereby advance their careers. The strategy works reasonably well at Yanagicho, since team members are required to return to the units from where they came, if they came from other factories or divisions in Toshiba. This spin on the Prodigal Son story practically guarantees that they will work extremely hard. No one wants to return home a failure.

Getting Started

The first bilateral meetings between Toshiba and Visa International to negotiate the direction of the SuperSmart card development project occurred on April 24, 1986, more than three months after the project was officially launched within Toshiba.[16] The project was laid out in terms of four phases: an *exploratory* phase for considering various technical solutions for producing SuperSmart cards; an *experimental* phase to work up those preliminary solutions; a *preproduction* phase for refining design, component, and subsystem parameters; and, a *production* phase. Investment costs in plant and equipment during the first six months, the exploratory phase, amounted to $2.1 million.[17]

As Toshiba and its vendors dove into exploratory development work in Japan, Mr. Pino Franchini committed Visa International to do marketing research and to performing the following tasks:

- define the application of the SuperSmart card;
- estimate the market size;
- quantify the market potential for the card;
- create a market research simulation of card applications in a microcomputer environment;
- conduct market research on issues of card application and validation;
- support experimental applications of SuperSmart card technology among member banks;
- provide ongoing visibility for the project at trade shows, company conferences, and publicity events;
- issue 6,000 cards by March 1989 for market testing in the United States;
- mobilize Visa International's sales network for marketing the card—some 28,000 member banks, 6 million merchants and 160 million cardholders.[18]

Obviously, Visa International's role was a major if murky part of the project. What to make, the marketing problem, defines *when* and *how well* to make as well as *how many* cards to make. Answers to these questions determine the amount and kind of resources to commit and the times at which they will be committed. But the parameters of the exploratory and experimental phases of the project were never fixed because throughout 1988 and even into 1989 what exactly was being made remained unclear. Indeed, even as the engineering sample was being readied, some key marketing issues were unresolved.

Visa International had to make up its mind. It needed to conduct research on the size of the smart card market, the technical sophistication of end-users, and various ways of segmenting the market. Was the card to be an ultra-high-end application, a platinum card for the super rich, or targeted at the merely well-off? Given the overall size and diversity of Visa International's client base in the mid-1980s, it was imperative to determine exactly who was being targeted as smart card customers.

In the meantime, Toshiba was working on incorporating the card's technical requirements in a roughed-out prototype, called a "functional breadboard" (FBB). Even before the FBB stage, card functions were laid out in a software program to familiarize the team with the size, complexity, and content of Visa's smart card proposal. An FBB's physical dimensions are not at all important; at this stage, the goal is to construct a working model of the product, regardless of its physical size or appearance.

Since technical and functional requirements are intimately related to market requirements, the sooner that Visa made up its mind, the sooner Toshiba could refine the smart card features. So, when team members met at Yanagicho in April they set FBB requirements based on what they knew at the time.[19] With some initial objectives at last in hand, designers and developers buckled down to work. As their numbers swelled, the team was divided into two groups. Initially, one team of some 60 engineers tackled hardware issues and another 20 or so looked after software, including the on-line and self-contained software features of the card. Since Toshiba already had experience with on-line software development for the Mitsui Bank card, 20 software engineers seemed sufficient at first.

But numbers climbed rapidly as design work advanced and as FBB work was completed in January-February 1987. During the first five-month FBB phase, 16 strategic planning meetings were held at Yanagicho covering a diverse range of topics, including NTT's (Nippon Telegraph & Telephone) on-line software protocol system, efforts to integrate semiconductor-based functions onto one chip, International Standards Office (ISO) standardization, and key component and circuit reliability.[20] Six months after launching the FBB phase, in October 1987, Yanagicho laid out what had been accomplished in three areas, hardware, software, and documentation (see Table 5.1). Once FBB features tested out, the next phase, the preproduction phase of the project, was scheduled with a goal of building an actual prototype of the world's first credit-card-sized computer.

Table 5.1 Functional Breadboard Requirements and Number
of Development Units for SuperSmart Card Project

Classification	Item	Quantity
Hardware	(1) functional breadboard with keyboard/display card, transducer card, and control card	2 sets
	(2) spare card with keyboard/ display card, and transducer card	1 piece 1 piece
	(3) transducer reader	1 set
Software	Card Holder Interface Software	1 set
	(4) idle state	
	(5) time and date state	
	(6) account functions	
	(7) calculator	
	(8) ancillary functions	
Documents	(9) operations manual for FBB	2 copies
	(10) operations manual for transducer reader	2 copies

Product Development Team Design

The SuperSmart card project boasted some 172 management, design, and development members at its zenith. Not all of these were working full-time on the card, of course, but even so, carrying nearly 200 high-priced engineers and technicians on the smart card budget gives an indication of how important the project was to Toshiba. [21] The organization of the project and the number of Yanagicho employees, as opposed to other Toshiba and non-Toshiba personnel, in core development and design activities are shown in Figure 5.2.

The large numbers of corporate level personnel associated with the project, 47 of out 172, also suggests how important the project was for Toshiba. Top-level managers and scientists from key staff functions and research facilities lined up behind the smart card concept. Obviously, their involvement in the day-to-day development of the card was minimal but their high-level support was significant not only for the key personnel and resources that they routed to Yanagicho but also for the aura of legitimacy that they lent to the project.

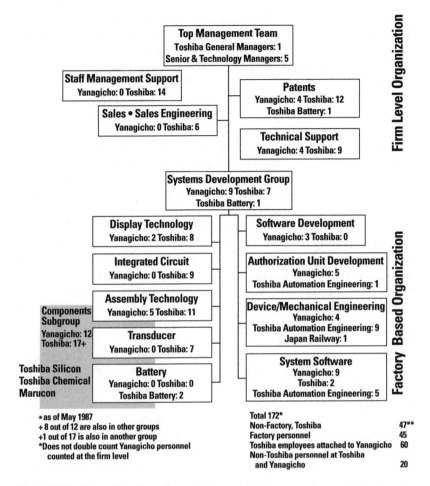

Figure 5.2   SuperSmart card project organization.

Forty five members (or 26 percent) of the project team came from Yanagicho. In other words, one-quarter of the engineers and technicians were from Yanagicho, even though Yanagicho, was the project's one and only development platform. This confirms one of this study's key findings: both strength and flexibility are accommodated within Knowledge Works. Yanagicho has the ability to pioneer and promote new development activity while, at the same time, it has the flexibility to absorb researchers and technicians from elsewhere and follow their lead. Of the ten factory-level, project subgroups, five groups were especially dominated by Yanagicho personnel, 3 to 2: systems development, software, systems software, the precision machining-oriented authorization unit, and device/mechanical engineering teams had 30 Yanagicho personnel out of a total of 56. Yanagicho's strategic position in the project is especially evident when it is noted that 16 of the 26 non-Yanagicho personnel working in these areas actually came from two companies that were more closely affiliated with Yanagicho than with Toshiba itself.

Toshiba Automation Engineering is a Toshiba-affiliated company that serves as a software supplier to the Yanagicho Works. Virtually all of its software development activity is captured by Yanagicho. Japan Railway is a privatized offshoot of the former Japan National Railway (JNR) Company. Presumably one of their employees was seconded to Yanagicho because of the close working relationship that had evolved between the factory and the company during the many years that Yanagicho was a key supplier of automatic railroad ticket issuing and validating equipment to the JNR.

While Yanagicho personnel were clearly central in software development and in the hardware areas of designing authorization units and reader/writers, they were not so evident in the design and development of smart card internal parts and components. Here, Toshiba personnel from other parts of the company that specialized in display technologies, integrated circuit design, and transducers were quite evident as well as non-Toshiba personnel from affiliated companies, like Toshiba Battery, Toshiba Silicon, and Toshiba Chemical. The only subarea in which Yanagicho personnel were significant was in assembly technologies. While the leader and subleaders of this group were from Yanagicho, the core members of the group came from the central R&D lab and Toshiba's specialized units for new materials research, new manufacturing technologies, semiconductor substrate design, and semiconductor manufacture.

Overall, personnel from 10 Toshiba divisions and research units as well as from 5 affiliated and 2 nonaffiliated companies participated in the project. Employees from the Semiconductor Division and the Ultra-Large Scale Integration Research unit worked on the design and development of the central processing unit (CPU), the memory unit, the LCD driver, and the reliability of these units when running alone and in tandem. They sought to integrate more and more functions on fewer and fewer chips, ultimately seeking one chip to drive all microprocessing and memory functions. From the Audio Video Research Lab came specialists in transducers, and from Electronic Devices came LCD display engineers.[22] Toshiba Battery brought its ultrathin battery technology to the party, while Toshiba Silicon and Marucon helped develop the calculator-style, touch-pad keyboard. Toshiba Chemical contributed embossing know-how that proved critical to the card's manufacturing process. Obviously, close cooperation between Yanagicho and other Toshiba and non-Toshiba units was a key feature of the smart card development process.

The importance of Yanagicho personnel for the design, development, and assembly of smart cards epitomizes the place of Knowledge Works for integrating industrial R & D and manufacturing functions for Toshiba. Yanagicho was licensed to innovate across a broad range of technologies and functions, and having done so, the accumulation of systems knowledge and all of its future applications, adaptations, and developments accrued to Yanagicho. Hence, the decision to give the smart card project to Yanagicho as opposed to Ome or some other Toshiba factory was a momentous one. The lion's share of the benefits and, of course, most of the blame, would rest squarely on Yanagicho's athletic shoulders.

SuperSmart Card Prototype

Software and hardware problems bedeviled the project from the start. Yanagicho engineers submitted a master list of 16 problem areas that still plagued the project 15 months after the official launch.[23] The most severe were in four key areas: the size of the software program to be crammed onto a 16K chip; and the design of a two-way card that could be used either with sliding imprint authorization units, the traditional technology, or with ATMs, the rapidly emerging authorization technology. Equally important was the improvement of the overall reliability of the LCD display and key components of cards, especially when they were carried in hip pockets and sat on by male users; and the power and size of the battery unit driving most of the other key components.

Miniaturization and reliability were the main headaches and, not so surprisingly, they were inversely related. The smaller the physical dimensions of the card, the more problematic were heat dissipation, strengthening the LCD area, counteracting LCD bleeding when using mechanical authorization units, and preventing sudden changes in ambient temperature and humidity from penetrating into the LCD package area. The greater the efforts to miniaturize, in effect, the more likely mechanical, electrical, and physical failures.

In April 1987, when the first prototype cards appeared, the failure rate was 90 percent, that is, only one out of ten cards was testing properly. With less than four months to go before Visa took delivery of 50 cards for a two-year test period, the big question was whether or not 50 reliable cards could be assembled! To reach that target, prototype cards were put through a series of 19 different reliability tests. These included static electricity, bending or torsion properties, temperature and humidity, durability, resistance to chemicals, and electromagnetic field resistance. Passing one of these tests, unfortunately, was often at the expense of another.

In August 1987, with prototype cards well into the engineering development stage, Yanagicho initiated a series of tests concerning the card's physical dimensions and characteristics, and its reliability under various instances of environmental stress. While a millimeter here or there, say, in LCD dimensions or semiconductor contact points, seems fairly unimportant, an oversized card would not function properly in ATMs and card authorization units. Fortunately by August 1987, the prototype card checked out in six key areas, including its ability to dissipate heat and resist external electrical charges.

But the most taxing checks were tests of environmental stress. There were 19 different checks, including static electricity endurance; torsion and bending properties; low- and high-temperature operations; low- and high-humidity operations; battery life; resistance to chemicals, especially alcohol and gasoline; and X-ray and electromagnetic field resistance. While the card passed easily in most areas, sometimes the card would fail miserably, such as resistance to spilled gasoline. Since gasoline station sales are one of the most common credit card uses, card failures like this could jeopardize the entire development program. By September 1987—later than planned but within an acceptable timeline—the card passed both hardware and software reliability tests.

Notwithstanding these difficulties, Yanagicho engineers committed themselves to delivering prototype cards, card readers, and card authorization/personalization units by late September 1987. This was just a few short weeks after a master list of problems had been discussed in a bilateral meeting in San Mateo, California. Yanagicho had insisted that the cards, card readers, and authorizations units would be made available first in Japan, not in California. And, importantly for the future of the project, two different sorts of cards were to be delivered: one a touch display (TD) type and another a magnetic stripe (MS) type. Yanagicho had not yet solved how to combine the best features of both cards in one prototype.[24]

The next bilateral meeting took place at Yanagicho on September 24. A prototype card was ready. Amazingly, it was a combined TD and MS card. It boasted on-line credit authorization and approval code generation features with an internal algorithm to check for code accuracy; a front panel with embossed characters and hologram design for compatibility with traditional cards; an LCD that had been strengthened against bleeding; an internal clock with a monthly error of less than two minutes; low battery detection; and a CPU that did not lock up when the keys were punched quickly. Yanagicho was proud of what it had accomplished. Figure 5.3 reveals how far Yanagicho's efforts were in advance of the conventional technology.

But Visa International was not satisfied. Visa felt that the amount

Components	Items	SuperSmart Card	Conventional Technology
LSI	Number of Chips	2 to 3 chips	LSI 8 chips MSI,SSI 80 chips
	Supply Voltage Power Supply Current	2.2-5V	5V
	•Operating	0.5mA	22mA
	•Standby	1uA	10uA
Liquid Crystal Display	Digits Font	16 Character 5x7 dot matrix Alphanumeric	8 Digits 7 Segments Numeric
Crystal Oscillator	Frequency Thickness	32.768 Khz ± 15 ppm (at 25°C) 0.5 mm	32.768 Khz ± 20 ppm (at 25°C) 1.0 mm
Battery	Size Voltage/Capacity	24x38x0.5mm³ 3V/80 mA	16x34x0.5mm³ 3V/20 mA
Transducer	Function Thickness	Magnetic Stripe Emulation 0.3 mm	

Figure 5.3   Main components of the SuperSmart card.

and angle of force to activate the touch keys were unacceptable; the internal contact points would have to be altered. Also, the response time to keyboard entry was too slow, undoubtedly because the software program was large for the amount of available RAM. Finally, the card was a mite too thick and its torsion resistance inadequate, or so said Visa. Visa seemed intent on getting it absolutely right.

Improvements to the card after September 1987 were targeted in four different areas: the sensitivity of the keyboard; LCD bleeding caused by high temperature and mechanical stress; stability of the crystal oscillator; and assembly process yield. In each area, Yanagicho team members had possible solutions. First, they enlarged the contact points for the keyboard and improved the carbon contact silk screen technology. Second, they made the LCD seal area larger and improved its barrier coating. Third, they enhanced the stability of the crystal oscillator by tightening the mounting for the tuning fork and improving the CPU oscillator circuit. Finally, by insulating the materials used between components, the manufacturing yield improved. Laser welding, it appeared, advanced the quality and reliability of the heat seal process.[25]

Test marketing of the SuperSmart card prototype was mostly carried out in Japan, although *Time* magazine ran a very upbeat article on using the card at Camp Lejeune, a Marine Corps base in North Carolina. The testing group in Japan included Visa International,

Toshiba, NTT Corporation, and Visa card issuers. Because Visa's IC cards were designed to be compatible with existing reader/writer terminals, virtually any place in the world could be used for testing purposes. In retrospect, it might have been preferable to have done more testing overseas because the card's fancy encryption algorithms were not given a real workout in Japan where credit card fraud is minimal.

Something much worse than credit card fraud jeopardized Yanagicho's development efforts, however. Late in 1987 the "Toshiba machine incident" blackened Toshiba's reputation among the nations of the West, threatening the company's products in numerous Western markets. Toshiba Machine, a 51-percent-owned subsidiary of the Toshiba Corporation, was fingered as the supplier of numerically controlled, multiaxis machine tools to the Soviet Union. Matched with Norwegian software, Toshiba Machine's NC tools turned out propellers for Sovier attack submarines that were much quieter than previous propellers. Sales of such equipment were expressly banned by COCOM, a regulatory body of cooperating industrial nations organized to restrict or impede the sale of strategically sensitive materials to communist countries.

Toshiba reacted quickly with a well-oiled lobbying campaign in Washington and a comprehensive COCOM compliance program to prevent recurrent violations. The compliance effort affected the testing of the SuperSmart card. Toshiba wanted assurances from Visa that it would not export strategic materials related to the card's design, especially the double-sided bonding, thin-film battery, and reader/writer technologies, without Department of Commerce authorization.

For its part, Visa International was reluctant to guarantee DOC authorization for every export of SuperSmart card technology since Visa would not be able to control how the card's technologies would be used. In fact, Visa International was not a single integrated corporation but a federation of national card-using merchant organizations. Visa might be able to document how many units would be sold to whom but it was in no position to monitor how SuperSmart cards were actually used in the field. The testing program slowed in the aftermath of the Toshiba machine incident, especially in the big city markets of Western Europe and North America. Limited ATM tests at the Great America First Savings Bank in San Diego and National City, California, were not held until the end of February 1988.

Reliability Problems

The biggest technical bottleneck was the thickness of the card. Making a thinner card was not such a problem, but making it thinner and not creating other problems was. The thinner the card, the less internal clearance for the LCD display, wire bonding connections, IC, and the transducer. And any one of the these, but especially the LCD display, could be damaged by the grinding and abrasion experienced during normal use. Since the card was designed to be used with sliding imprinters in the traditional way, embossed numbers had to be added to the exterior dimensions of the card. The internal thickness of cards used in imprinters was 1.12 to 1.14 mm thick and, on average, the embossed portion was anywhere from 0.20 to 0.36 mm thick.

In trying to solve the trade-off between the card's internal and external dimensions, Yanagicho submitted two sample cards to Visa International, one of 0.995 mm and another of 0.900 mm thickness. The difference between the two was not just 0.095 mm but also the embossing method and mounting. No more than 0.1 mm or so was at issue, but 0.1 mm can make all the difference in the world. Manufacturing a semiconductor chip of more than 0.1 mm is routine; making one less than 0.1 mm is not. In a 1989 memo from Visa International to Kouji Kuramochi, a divisional manager at Toshiba, the following advice was offered:

> After they [plastic cards] are used in the imprinters, and are carried in wallets for a while, they usually have an average thickness of 1.12 to 1.14 mm. Therefore, any cards that have overall thickness of over 1.22 mm are potential problems. This is even more so for the older imprinters, which tend to "wobble" more, and may cause more difficulty with imprinting and may even cause the card unit to jam. . . .
>
> Since SuperSmart cards are engraved from a "hard" layer of adhesive material as compared to the regular plastic card's embossing, it is likely that they need not have the same height (as the embossed characters on plastic cards), and still give good imprints. Therefore, in theory, an overall thickness of less than 1.22 mm can be accomplished by reducing the engraving height of the 0.900 mm card from 0.48mm to 0.36 mm, and for the 0.995 mm card's engraving height to be reduced from 0.43 mm to 0.20 mm. However, in reality, the quality of the imprints produced from the reduced engraving has yet to be tested.[26]

Yanagicho soldiered on. Two months later, a May progress report read, "During a test of cards among employees, a 50 percent card failure rate (28 cards) was experienced." Difficulties continued with crystal circuit failures, cracks in the LSI, LCD bleeding, and smudging of the imprint readings. Optical character recognition (OCR) scanning of sales drafts using mock-up cards in May found that cards of 0.900 mm thickness experienced a 10 percent rate of failure. (The acceptable read failure rate for embossed plastic cards was 15 percent.) The ISO specifications for credit card thickness was 0.76mm ± 0.08 mm, so Toshiba had to slim down another 7–8 percent without damaging card reliability. The same May communication noted that the Bank of Nova Scotia and BANESTO (Banco Espanol de Credito) were interested in test marketing improved SuperSmart cards.[27]

Incorporating magnetic stripe technology with the SuperSmart card was a particular area of difficulty. The addition of a magnetic stripe and embossing would allow SuperSmart cards to be used in traditional mechanical card readers, and hence retailers would be less resistant to using them. Smart cards could be introduced widely yet naturally without creating problems for retailers who, perhaps more than customers, would decide the fate of smart cards. Magnetic stripe technology, however, forced a number of problems to emerge. Stainless steel proved to be a difficult host metal for mounting the magnetic stripe. A low magnetic permeability stainless steel had to be used, and a special layer of unusually receptive material laid down first in order for the magnetic stripes to adhere properly. Finally, because of LCD bleeding caused by mechanical and torsional pressures, a partially flexible card with high-endurance LCD was developed. By the end of the project in fact, a fully flexible stainless steel card was being manufactured.

More Reliability Problems

While the initial cards were undergoing field testing, Yanagicho's development team continued to improve the card's key components and overall reliability into the first half of 1988. Ongoing improvement was important for several reasons. Most important, as Visa International's market testing program advanced, new areas were identified for improving the card. Designing and developing the right card, not just any card, depended on market feedback. Second, reliability problems still bedeviled the card and it was desirable to keep the team occupied and motivated with these during the long testing period.

While most of the redesign effort focused on hardware, software testing and debugging were other areas of ongoing development activity. In fact, the software side of the house ran two development projects in parallel, important in case one approach fails. One team loaded operational data in RAM (random access memory) while basic and optional software supporting the card's operation was in ROM (read-only memory). A second team put basic program software into ROM but stored optional software features and operational data in EEPROM (electronically erasable and programmable read-only memory).

The EEPROM approach was superior in ease of use, reprogramability, and protection of software from malfunction, but it required modification of LSI (large-scale integration) chips and the software program, including some rewriting of the software protection code in the LSI program. The merits of the RAM-ROM approach, on the other hand, were twofold. Current LSI chips could be used without modification, and the card's data storage capacity was relatively greater. However, lead times for changing software specifications were longer, and once cards were in the customers' hands, changing specs would require a high level of service capability on the part of field representatives. The two teams worked simultaneously throughout 1988.[28]

The hardware redesign effort concentrated on three key component areas of the card, the LCD, the crystal oscillator, and the ultrathin battery. LCD problems arose in several ways. Air bubbles would appear under the sealed area during manufacture and render the unit useless, and ATM mechanical and torsional pressures would break the adhesive barrier protecting the LCD, destroying the unit. The area of the seal was enlarged and its adhesive material improved as a result. But there were physical limits to the size of the seal—the obvious solution to the problem of air bubbles—given that all the components and circuits were crammed into a space smaller than a plastic, magnetic stripe credit card.

The crystal oscillator was another problem area. When the card was dropped, sat on, bent, or run through an embossed code reader, the oscillator often failed. As always, in this instance as others, the solution was to increase the package thickness protecting the tuning fork from 0.5 mm to 0.55 mm, and the tuning fork mounting was strengthened by increasing the cavity housing mass. By the spring of 1988, the crystal oscillator's reliability under various adverse conditions improved remarkably while the card's thickness increased by no more than 0.05 mm or about the width of a human hair.

The main difficulties with the ultrathin battery, as with most of the precision components, were the mechanical and torsional pressures exerted by ATMs and card imprinters. Thickening and thereby strengthening the component area were the solutions although, as always, such solutions impinged on the critical space needed to protect other components and circuits. Rather than increasing the mass of the battery itself, the edges of the compartment used for housing the battery were reinforced. Finally, using a defocused laser, battery terminal connections to the PCB were measurably strengthened.

The assembly process also advanced under the able direction of Kazuichi Wada, a design engineer collared from Yanagicho's Manufacturing Engineering Department. Wada's responsibilities—the manufacturing process and automatic assembly technologies—were the keys to the entire project: if high-performing, high-quality cards could not be made reliably and inexpensively, then the project was a bust. Under Wada, the assembly process progressed in three areas: the card's layout and parts count were redesigned to simplify assembly, thereby lowering costs and increasing reliability; a wire bonding process was developed for the CPU, LCD driver, antisurge device, and memory chip, which by 1988 was upgraded to 64K; a two-sided, through-hole, high-density printed PCB assembly process was perfected. The PCB assembly held the condenser, the chip shield, contact points for external readers, and two crystal oscillators of 200 KHz and 32.8 KHz.

Single-fiber gold wires from 110 contact points were used to connect the chips and LCD driver, welded by a defocused laser beam. The wires were low loop, through-hole in design and encased in epoxy resin. Epoxy resin does not expand under heat, although the lead frame in the resin will expand if enough heat penetrates the resin casing. The two-sided, through-hold PCB was sourced from the Seiko Company, but Wada's job was to design an assembly process that ensured proper functioning for years to come. Wada succeeded admirably, as witnessed by the extremely low rates of failure for the wire bonding process and hardened PCB.

Manufacturing assembly technologies underwent a monthly design review. At these meetings, suggestions regarding product/ process innovations were put forward; advances since the last meeting were reviewed; budget targets were assigned and scrutinized; and future directions were aired. The autonomy of teams to try their hand at novel solutions was assumed, although it was never clear who bore the responsibility for failures. Wada was responsible for whatever happened in his section, although he could not keep track of everything his teams were doing. Wada encouraged tinkering and experimenting with the production layout and assembly process, knowing full well that he alone was responsible for any failures.

By early summer 1988, the assembly teams' labors were paying off. Out of an engineering sample of 222 cards, the failure rate was down to 3 percent, a drop of some 35 points from a year earlier. Most important, version 1.62 of the card experienced no failures due to LCD bleeding or battery disconnections.[29] Even so, problems with circuit reliability and the crystal oscillator continued to plague the card; increasing card rigidity on one hand led to a decline in flexibility on the other. The reality of development work is hundreds of trial and error experiments, guided by some notion or intuition as to what might work better. Small numbers of these experiments actually work out for the better.

Notwithstanding the difficulties of development, Toshiba's confidence with its development team resulted in a six-year projection for capital investment and production quantity in September 1988

Table 5.2   SuperSmart Card Investment and Expected Quantity*

	1986	1987	1988	1989	1990	1991
R & D investment (million$)	6.6	5.2	4.7	2.5	1.2	0.8
Production investment (" ")	1.8	2.1	0.3	1.0	4.4	3.0
Full-time Engineers	42	46	41	22	15	15
Production Volume (1000s)	—	—	15	100	1000	5000
Toshiba Market Share (%)		100	100	100	50	33
Unit Price (1000 yen; Toshiba FOB)			11.0	4.6	3.2	2.2
Unit Price (US$)			42.10	13.48	9.27	8.42

*These cost and volume estimates come from three memos found at the San Mateo, California, offices of Visa International, dated March 2, 1988 and September 1, 1988.

(see Table 5.2). Though the projections were made for Visa International's benefit and cannot be considered completely reliable, they do suggest something of the amount of investment that Toshiba was putting into the project.

Toshiba *valued* what was invested to date at more than $20 million dollars in 1988. That occurred overwhelmingly at Yanagicho. This underscores the importance of Yanagicho as a product development platform and highlights the role of Knowledge Works in Toshiba's manufacturing strategy. Toshiba cannot sell what it cannot make, and it cannot make what it cannot design and develop in a timely and cost-effective manner.

By the spring of 1989, cards of high reliability, capacity, and performance were being produced.Computer analysis of the card's structure, components, and circuitry was yielding additional knowledge of how to enhance the card's reliability and performance. By using thin panel materials, the thickness of the PCB, the crystal oscillator, and the mounting adhesive were further reduced. Also, by introducing slotted groves into areas surrounding key components, stress conduction was greatly ameliorated. Only 2–3 percent of the cards used in field tests were failing, and they were failing under conditions of extreme environmental stress. If cards fished out of Puget Sound failed or if LCDs bled when the family station wagon ran over them, no one could really be blamed.

From early 1986 to the end of 1987, a 22- to 24-month period of intense development activity, a spiraling path of progress from the initial FBB stage until delivery of fully field tested cards to Visa International can be described. Most of the technical highlights of that journey according to a fourfold stage scheme are captured in Figure 5.4. Numbered boxes identify where in the value chain innovations and technical breakthroughs occurred.

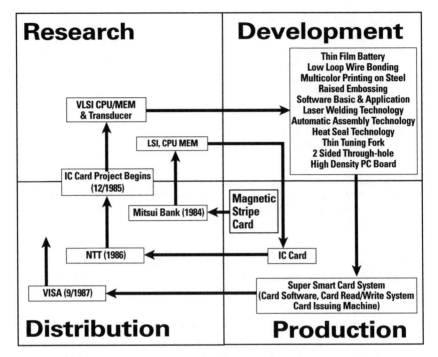

Figure 5.4   SuperSmart card project history.

Denouement and Project Redefinition

Just when Toshiba was finally mastering the intricacies of designing, developing, and assembling a smart card to Visa International's specifications, Visa International was getting tentative on how to market and promote the card. On January 11, 1989, Gretchen McCoy, a California-trained MBA who was being given more and more responsibility for the project on Visa's end, wrote to Mr. Mitsuo Kubo, general manager of Toshiba's Business Machine Division, that Visa International wanted to reposition the card as a "niche card" for the travel and entertainment market.[30] Shortly thereafter, on February 14, 1989, three years and a few months after the smart card project was launched, Mr. Charles Russell, president of Visa International, further startled Mr. Kubo, with the following letter:

> First, early outcome of deregulation has had quite the opposite affect, causing telecommunication costs to decrease significantly. This situation provided Visa Members an opportunity to implement cost-effective on-line risk management techniques to control a greater portion of credit and fraud losses. Secondly, member bank enthusiasm for the IC Card technology quickly subsided at the end of 1987 due to unfavorable IC Card economic assessments.[31]

In short, card costs were excessive and according to Visa International marketing surveys, the self-authorization features of the card were not perceived as a convenience. Also, merchants were reluctant to invest in new point-of-sale equipment that would allow full expression of the card's off-line features.

In short, Toshiba was snookered. The revolutionary on-line features of the card were judged inopportune while the off-line capabilities required more investment to be used with existing equipment. Moreover, Visa International was narrowing its focus to a specialized market niche, the travel and entertainment market. The mass appeal of the card—self authorization for users and compatibility with existing equipment—had all but disappeared in Visa's loss of marketing and strategic nerve.

Russell's letter continued:

> These events lead to the evolution of the "Flagship of the Brand" concept. As you know, initially we plan to test this concept with the travel and entertainment heavy user segment. We feel that the card's recognition value, distinctiveness, and unique capabilities to store and retrieve information can be utilized as part of a complete service offering to target *this small*, but profitable and ever growing segment of consumers.

Russell promised that Visa would continue with research to validate the "Flagship of the Brand" concept, but his February 1989 letter, coming some three years after the project kickoff, effectively killed further development of Visa's SuperSmart card at Yanagicho.

Pot Spin-Off

While the Visa project floundered, Yanagicho and Toshiba developed a series of new products based on innovative technical features of the SuperSmart card. The development team that created smart cards was folded into a new IC Card Systems Department, and the first volume IC card products began rolling off the Yanagicho line in late 1989. If you buy a ticket on board the Hikari Super Express that practically flies between Tokyo and Osaka, the conductor calculates your fare and prints your ticket on a HandiTerminal, one of the SuperSmart card's follow-on products. And if your car is stolen in eastern Japan, most policemen carry a Toshiba POT (Portable Terminal) on the beat, a handheld device carrying a daily "hot sheet" of 20,000 stolen vehicles, another spin-off from the smart card effort.

In fact, the POT project is a good illustration of how economies of scope associated with product development can deepen and broaden capabilities across a spectrum of new products and processes. The POT technology was leveraged directly from the SuperSmart card project—a 16K CPU ASIC (application specific integrated circuit) chip, 768K memory chip, back-lite LCD, removable cartridge battery, and 31-touchpad keyboard—and in August 1987, less than a year from initial discussions, an engineering sample was delivered to the Kanagawa Prefectural Police Department. Compared with the portable terminals previously carried by the police, Yanagicho's POT was 40 percent smaller in size and 20 percent more functionally versatile. The processing speed of the new terminal unit was eight times faster, plus there was the convenience of data storage and retrieval by IC card rather than by floppy disk. The components came directly, almost without modification, from the smart card project, except that the keyboard and software were redesigned. The software code was now COBOL and Lattice-C running in support of MS-DOS.

MYSTEP was another, equally successful, spinoff. MYSTEP is an electronic notepad or personal assistant designed with engineering and technical applications in mind. Indeed, MYSTEP was designed coincident with the STEP 90 organizational campaign (organizational campaigning was discussed in chapter 3), and its use in STEP 90 allowed for closer coordination between resource allocation planning and on-the-job use of resources. Less slack on the job saves time and money, resulting in a tighter fit between individual and organizational goals. POT hardware was also used for MYSTEP, lowering manufacturing costs by generating larger volumes.

The transfer of IC card technology from Visa's SuperSmart card to POT and MYSTEP did not happen automatically or mysteriously. People transfer technology. And because people transfer technology, Toshiba has adopted a system of career advancement for technical specialists. Called the *Gishicho-seido* (Technology Specialist System), Toshiba has a parallel track for technologists that runs from the very top of the firm down to factories. Independent of managers who run business units (BU), *gishicho* concentrate on effective development of technologies, including such matters as standardization, reliability, and safety, as well as the allocation and education of technologists, the engineers and technicians that work for Toshiba.

At every step of the operational hierarchy, *gishicho* are there in consultation with BU managers, making sure that Toshiba technologies are not sacrificed to short-term business interests. In 1988, there were five *gijutsu shikan* (technology officers) attached to Yanagicho under the Technology Specialist System. At the height of development for the SuperSmart card, POT, and MYSTEP, four out of the five were attached directly to IC card projects. Given that people transfer technology, Toshiba was making sure that IC card technology was effectively generated, transferred, and applied at Yanagicho with the Technology Specialist System.[32]

Smartcards: A Smart Idea For Toshiba?

People do not move easily between large firms in Japan, and so there is little doubt as to the value of investing in them. But unless employees are properly motivated and managed, company performance will inevitably decline because regular employees cannot be laid off, dismissed, or arbitrarily transferred. Even in the midst of the latest recession, the worst of the postwar era, the number of regular employees dismissed from large firms remains small. Contrast this with the downsizing epidemic in the United States where, for example, Boeing reduced its employees by 40,000 since 1990 and in January 1996 AT&T laid off 40,000 in one fell swoop.[33]

"The company is the people" is an oft-quoted aphorism in Japan, and it rings true in the SuperSmart card case. Thus, as familiarity grows among project members and with suppliers, project component technology, manufacturing processes, product specifications, and project management skills can be expected to improve. Familiarity and effort paid off in the technical wizardry of the SuperSmart card and in the development of a host of follow-on products, like POT, MYSTEP, and HandiTerminal.

But familiarity by itself is insufficient. Successful team building and functional integration were all-important in overcoming the technical, engineering, and manufacturing obstacles that impeded the smart card's development. Participation in R & D projects is one of the most important ways to differentiate oneself from a couple of hundred other, equally talented technicians and engineers in Japan's large firms. Employees from Himeiji, Horikawacho, Yokohama, and Tamagawa plants plus non-Toshiba employees from Seiko, NTT, Mitsui Bank, Marucon, and Toshiba Battery were involved in the smart card and other IC card system projects.

Since successful participation in development projects is crucial to career success in Japan, there is little reason to fritter away time and energy on alternative activities. Get onto good projects, do a good job, and hopefully return home a hero.

The significance of building a reputation for delivering the goods in tightly stretched situations is captured in the term *kao passu*, or literally "face pass".[34] At Yanagicho it signifies a problem solver, someone with the reputation for getting the job done, and someone to whom you can confidently delegate responsibility. Such individuals, like Tamada and Wada, cement and catalyze human relations at Knowledge Works: they lead others in the charge up the development hill.

Biases in favor of employees sticking it out and seeing it through give Japan's firms strong incentives to build internal capabilities and competencies. Indeed, wages are far more tied to ability than seniority in a recent survey of 515 firms, reaffirming the current emphasis on talent, effort, and teamwork in contemporary compensation schemes.[35] Systems integration, in the fullest sense of the word—social, technical, and organizational integration—of specialized human, material, and intangible resources is the overarching goal of management. And in this goal, building resources for the future takes precedence over exploiting current resources. The ease with which resources, human and otherwise, move in Western firms may be, ironically enough, the weak link in their management processes and product development capabilities.

Smart Card Retrospective

The most demanding part of designing and developing the Super-Smart card were marketing, not technical, problems. These were Visa International's, not Toshiba's, responsibilities. Yanagicho succeeded in making the card, making it to specifications that were beyond even Visa International's most stringent requirements. The unsolved issues were Visa's unwillingness to convert its retailers to a POS system with EFT capabilities and its inability to decide which market segment or segments to target with the card.[36]

Instead of redefining the market with its millions of plastic card users and hundreds of thousands of authorization units and ATMs, Visa International chose to extend the existing market and not to capitalize on a technology that could obsolesce the plastic cards of rivals while not damaging its own installed base of authorization units and ATMs. This was a grand failure of Visa's marketing strategy and vision. Stated less critically, Visa underestimated what it could do and wanted to do in the industry.

From a security point of view, for example, an IC card with input/output capabilities could greatly reduce security risks associated with access to televideo and telecommunication services. While these services are already available to plastic card users, sophisticated add-on security checks are not possible with plastic cards that lack internal processing and programming capabilities. IC cards could encode what was being watched, costs of viewing, and accumulate charges over a given period.

IC cards could also decode transmissions that were scrambled for security reasons. Cellular phone security could be enhanced by using insertable IC cards with PIN numbers and automatic billing features. Sophisticated card uses of this sort are just beginning to appear in the marketplace today, although Visa International could have had them three or four years ahead of the pack. While IC cards are fairly common, witness how widely available IC telephone cards are found in France, England, and Japan, high-performing IC cards with input-output functions and a programmable memory are not. Failure to exploit these features must be hung on Visa International's neck.

Cost of the cards was Toshiba's problem, however. As long as production volume was in the hundreds (true as recently as 1992), the manufactured cost of the cards was in the range of 3,000–4,000 yen each. Even when per unit costs dropped to 1,000 yen in 1993, SuperSmart cards were 10 times more costly than comparable plastic cards. Of course, IC cards offered a lot more for the money. But higher production volumes, sufficiently high to really lower manufacturing costs, will come only after various standardization problems are solved. Toshiba has been involved in ISO negotiations for 10 years trying to fix the dimensions of IC credit cards and the location of the contact points for the IC and other components.[37] Notably, Visa International never attended ISO meetings.

An ordinary credit card or cellular phone becomes a personalized product with features unavailable to anyone else by using a PIN. This is highly desirable today when high-volume, me-too products are all too common. Personalization encourages a dialogue between buyers and sellers that leads to even more individualized products and services. Tomorrow's smart cards will be akin to personal electronic servants, responding to the commands and whims of only one boss, you. In short, Visa International had a revolutionary product but a decidedly traditional vision. It was more interested in pushing the credit functions of existing cards, where revenues were guaranteed, than it was in providing customers with fail-safe EFT capabilities. Marketing plans for the card were flawed from the outset, so the convenience, security, and reliability features of the card were never really promoted, at least not within the development period set by Visa International.

In all fairness, however, recalcitrant retailers may have been as much of a problem as faulty marketing. Retailers were worried that they would have to invest in new POS terminals and perhaps in new authorization equipment to take advantage of the high-end features of the SuperSmart card. And they would have to train employees in the use of both. With other card companies not putting similar pressures on retailers, Visa was facing a critical test of its value-added marketing and strategic leadership. It was no test at all. Visa International's smart cards never really got beyond the prototype stage of development.

If they had, the results could have been spectacular: hundreds of millions saved in spurious and criminal credit card charges, customer billing at the point of sale, zero credit losses, and zero credit limits. But we live in a world of competing technologies, and with telephone charges dropping dramatically and with the number of value-added networks (VANs) increasing rapidly in the late 1980s and early 1990s, the SuperSmart card offered a technical solution that Visa International did not really know how to capitalize on.

Toshiba benefited from its relationship with Visa even if the SuperSmart card languished. Toshiba racked up a thousand patents in the project and many of these have continuing currency. Partly as a result, Toshiba registered the highest number of patents for any firm, foreign or domestic, in the United States in 1992. Also, Toshiba transferred smart card technology to ongoing development projects with NTT, the Ministry of Post and Communications, and Bull France, with whom Toshiba is developing a coinless, autodialing, IC telephone card. Another benefit of the project was the extension and enhancement of EEPROM technology for both single- and multipurpose cards, an area where Toshiba had not been strong previously. The same could be said of thin-wafer battery technology, IC, and other component packaging technology, and certain aspects of software engineering technology. All of these were considerably advanced by Toshiba's involvement with the SuperSmart card.

The value of the payoff and spillover from the project, given what Toshiba invested, seems about right. Direct benefits realized in most R & D projects are not great, yet the innovativeness of the SuperSmart Card, HandiTerminal, and POT were linked directly to Yanagicho's site- and relation-specific capabilities. Without the Knowledge Works architecture, a multifunction, multiproduct, and multifocal factory with R & D facilities next to the shop floor, it would have been impossible to resolve ISO, standardization, and reliability problems within a two-year period. Without a factory next door to a lab, however, the lab would not have known what kind of R & D to do in the first place.

Ultimately, Knowledge Works are more important for their potential, that is, for the range of all possible technical and organizational applications, rather than for what they actually produce. Actual products may prosper or not, witness Kanagawa Prefectural Police's POT and Visa's smart card. Actual products are only glimpses of Knowledge Works potential. Employee motivations, teamwork, career development, learning, training, and education—all of these are equally important aspects of Yanagicho's potential.

The strategic issue for Toshiba was not how many cards were produced at what cost and price but how to mobilize its systems integration powers in the interest of producing high-value-added, high-performance products with reliability and speed. So, in the end,

Visa's SuperSmart card was simply an occasion for Yanagicho to strut its stuff: to exercise its systems integration capabilities, plumb its knowledge base, and mobilize the power and integrity of its resources. Yanagicho was up to the challenge even while Visa International was not, thereby confirming in the process one of Toshiba's basic aphorisms—"market in, not product out."

SIX

Internationalizing Knowledge Works

*The locals don't want to assume responsibility, so they don't
involve themselves in the work. The extreme* ukemi *attitude,
up-and-down the line, among workers, engineers, and managers,
is the problem.*

A Japanese manager in a Toshiba manufacturing
subsidiary on the Pacific Rim 2/20/90

The factory, as opposed to the firm, as a principal vehicle for transferring organizational and technical knowledge has been ignored or,
at least, badly neglected. This has led to a misunderstanding of the
process by which interunit and interfunctional manufacturing capabilities are moved, and to an overemphasis on strategic rather than
operational aspects of the transfer. When capabilities are site- and relation-specific, as in Knowledge Works, the transfer is likely to be especially problematic, resulting in wasted efforts and opportunities.

Waste is just what Knowledge Works are designed to reduce, yet
the very reasons why they reduce waste at home may be jeopardized
overseas. If the model cannot be exported easily, and if Japan's industrial firms fail to create good jobs abroad in some rough proportion to their manufacturing exports, then political and economic
pressures on Japan will inevitably intensify. "Good jobs" are the research, design, development, engineering jobs that are so abundant
as well as sticky in the Knowledge Works model.

International Background

These issues are important and becoming more so, because Japanese
foreign direct investment (FDI) in Asia, North America, and Europe
is large (US $100 billion in FY 1990) and growing (increasing 40–50
percent per annum until 1991 when the Japanese economic "bub-
ble" burst). Even so, the proportion of Japan's total manufacturing
capacity located outside of Japan is only 9–10 percent, one-third to
one-half less than the proportion for American industry, suggesting
that a long wave of heightened Japanese FDI may be expected for
some time to come.[1]

154

This is especially true for Toshiba. In 1992, Toshiba's export ratio, that is, how much of its domestic production was sold overseas, was 25 percent. The other two general equipment producers' export ratios were lower, 16 percent for Hitachi and 18 percent for Mitsubishi Electric, suggesting how far Toshiba has to go in internationalizing production.[2] Moreover, given tough economic conditions at home and the surging value of the yen, FDI is likely to continue at high levels throughout the 1990s. Such conditions compel firms to rethink manufacturing organization and strategy.

Rethinking basic forms and what to transfer abroad are not things that Japan's industrial firms, including Toshiba, have spent much time on. Until now, almost all FDI earmarked for manufacturing went into downstream assembly, distribution, and sales functions, although the exact proportions spent for which activity varies considerably by industry and region.[3] Now, for a variety of reasons, that is changing; direct investments in R & D, design, testing, and engineering facilities by Japan's transnational companies (TNCs) are increasing in North America and Europe while, at the same time, sales of electronics equipment and components are falling at home, a 5.9 percent decrease and an estimated 3.3 percent decrease in 1993 and 1994 respectively.[4]

Reinventing Taro

This chapter explores the problems with taking Japanese advanced manufacturing practices overseas, including not only upstream functions (research, design, and development) but also downstream activities (production, distribution, and sales) and the all-important links between them. To the extent that either upstream or downstream operations are distinguished by site-specific, high-performance characteristics, transferability becomes problematic.

That is exactly what has happened at home since the 1950s. Factories have been differentiated into two sorts: those concerned primarily with a rapid, flexible, and innovative response to market needs; those that are focused on stabilizing production and driving costs down as functions of quality and volume. The strategies, structures, and site-specific capabilities of the two differ. This book is obviously concerned with Knowledge Works, the former sort.

Upstream, as opposed to downstream, investment is welcomed by host countries for providing high-value-added employment, but at the same time, exporting upstream activities is worrisome to Japan's TNCs for two reasons: first, it may lead to a hollowing-out (*kudouka*) of domestic industrial capacity; second, it may impair the linkage of upstream-downstream functions. So, while growing FDI in manufacturing seems inevitable, a resulting export of jobs, particularly upstream ones, worries Japanese firms and government officials. Also, increasing FDI may weaken certain home-grown advantages, such as a steady growth in market demand and in the size and quality of the workforce during the past four decades.

However, if Japanese TNCs wish to set up fully integrated operations overseas, which is consonant with the latest thinking about global business strategies, "good" jobs as well as "not so good" ones will have to go. TNCs must design and develop products as close to local markets and customers as possible. As this happens, critical strengths in Japan's just-in-time manufacturing techniques, fast-to-market product strategies, and multifunctional collaboration in upstream research, design, development, and engineering activities— the Knowledge Works model—will be impaired.

Overseas activities are likely to be hamstrung in critical ways. Even in Japan, as upstream functions have become increasingly important, it is clear that "mental" work as compared with manual work is hard to define, harder to regulate, and hardest to manage.[5] In Knowledge Works, employees have to be motivated beyond "doing a good job", but clearly specifying what is beyond "doing a good job" is no small task as it hinges on the context within which "goodness" is defined. Not surprisingly, exporting mental work into little understood cultural and industrial environments is a scary prospect.[6]

Knowledge Works: Problem or Solution?

Although charged with the central tasks of product and process innovation and market positioning for a huge portfolio of products, Knowledge Works are situated at the lowest levels of company organization. So while Toshiba is a multidivisional enterprise with divisions enjoying high levels of responsibility over product lines and market operations, in Yanagicho's case it answers to a dozen different business units, five different product divisions, three different R & D facilities, and numerous corporate-level staff offices. Clearly, the complexities of Yanagicho's work are not contained within any single division's domain, as illustrated in Figure 6.1.

In this, Toshiba's methods of factory organization and management differ from other electrical equipment/electronics manufacturers. Matsushita Electric Industry, for example, follows a strategy of extreme decentralization and specialization of manufacturing functions where, for the most part, one factory manufactures one product line for one division. In Matsushita's way of doing things, divisions are incorporated as independent companies, so that one factory manufactures one product line for one company. Hitachi, by

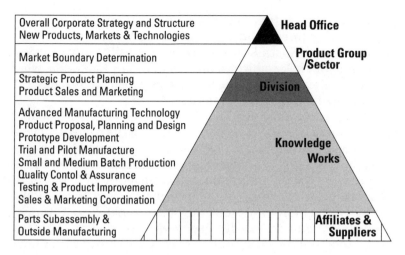

Figure 6.1 Corporate hierarchy and Knowledge Works.

contrast, is more structurally centralized than Matsushita because its multiproduct factories are their own profit centers.

Given Yanagicho's low center of gravity (but high centrality) in Toshiba's company organization, it is a bit surprising how prominently Yanagicho figures in the international manufacture of Toshiba plain paper copiers (PPCs). In the multidivisional model of the firm, the responsibility for overseas transfer of operations would normally fall to the divisional level of management. But PPCs (along with laser beam printers) are Yanagicho's most important product, representing about half of the total value of the factory's production. In recent years, moreover, as much as 50 percent of the production has been exported, mostly under original equipment manufacturer (OEM) contracts to 3M and Harris. (3M's distribution of photocopiers in North America moved to Lanier Worldwide in Atlanta, Georgia, in 1989.)

Moreover, PPCs are among a half-dozen office equipment products, including multi-purpose facsimile machines, laser beam printers (LBPs), microcomputers, and computer peripherals, that Japanese manufacturers are aggressively marketing and manufacturing overseas in order to supply products for the office of the future. Yanagicho's crucial knowledge of and experience with PPC design and manufacture—a product it has specialized in since 1966—has made it Toshiba's springboard for the overseas transfer of PPC technology.

Yanagicho first went abroad with the opening of Toshiba Systems France near Dieppe, in Normandy, France, in 1986, and in the same year with the acquisition of a 3M toner plant in Mitchell, South

Dakota. In the meantime, Yanagicho was instrumental in the implementation of three technology transfer agreements for PPC manufacture in India, South Korea, and China. Finally, early in 1989, a new facility for PPC production opened in Irvine, California, where Toshiba was already assembling laptop computers, fax machines, X-ray equipment, PBXs, and telephones.

So, even while Yanagicho has no branch factories in Japan, it boasts six branch factories abroad, attesting to its competencies in the design and manufacture of PPCs. However, when Yanagicho's factory-specific resources are juxtaposed with the loci of formal decision making for the international manufacture of PPCs (in the Office Equipment Division of the Information and Telecommunications Product Group), problems of a mismatch in authority/ responsibility with knowledge/experience are obvious. These are basic to understanding the difficulties of going abroad with Knowledge Works.

Mission without a Headquarters

These complications are nowhere better illustrated than at Yanagicho. Although two dozen personnel from a variety of departments, including PPC production, design, manufacturing engineering, production management, planning, and marketing, were working or preparing to work abroad from early 1986, they were unable to report problems, air grievances, and generally seek advice from a single, informed source until July 1, 1988. Nominally, the department head of manufacturing administration (*seisan kanri bucho*) was in charge of overseas operations but Yanagicho itself was more than enough for him to worry about. It was some 18 months after Toshiba Systems France had been founded and 3M's toner plant in South Dakota acquired that someone at Yanagicho was put in charge of overseas production.

Mitsuo Komai was a good choice for the job. He speaks reasonably good English and he has had a distinguished career at Toshiba and within Yanagicho. His specialty is manufacturing technology and he has worked in PPCs design and sales. For a while he was a technical consultant on Yanagicho's OEM contract to supply 3M with PPCs, and he has worked in PPC service operations as well. Before tackling his current assignment of overseas manufacturing, he was in charge of factorywide quality assurance and control, one of the most important staff functions.

While undoubtedly a good pick for the task of managing overseas operations, throughout the summer of 1988, Mr. Komai was without a budget, staff, office, or dedicated resources of any sort that would enable him to accomplish his task! While these deficiencies were corrected by 1989, until then Mr. Komai was simply attached to the factory manager's office and he enjoyed no managerial means or resources of his own. When I interviewed him in August 1988, Mr. Komai said that Toshiba's way of doing things is to force managers to take responsibility themselves, but throwing up his hands, he added that perhaps this was pushing the principle a bit too far.

According to Mr. Komai, the biggest problems with producing PPCs overseas are not manufacturing problems per se but the way of manufacturing or, even more generally, the way of thinking about manufacturing. "It's hard" he said, "to get non-Japanese employees to understand that their work is not just turning screwdrivers but worrying about total productivity."[7] It is not simply a problem that overseas workers do not want to work hard enough. In fact, he insists that French and American workers try as hard as their Japanese counterparts do. And it is not only a problem of inadequate documentation, in spite of the fact that not all of Yanagicho's manuals, instructions, and rules of proper manufacturing procedure have been written down and translated. Documentation and education are not sufficient in themselves, because while you can teach rules, you cannot teach a way of doing things. Moans Mr. Komai, "The nature of work is not in textbooks but in our minds! To really understand work rules, we should talk in Japanese with other Japanese." To understand the French or American way of doing things, Mr. Komai mused that he should spend a few months working and living with Toshiba's French and American employees. He would eat with them, work and relax with them, play with their children, talk with their husbands and wives. Finally, he would begin to understand them as people and appreciate their way of living and working. (In 1992, Komai got his chance to live in and work among Americans. Not too surprisingly, productivity and profits soared under his tutelage.)[8]

Komai's concerns and doubts about understanding the French and American ways of doing things are not trivial expressions of Japanese parochialism. Instead, they reflect a genuine appreciation of the extent to which Yanagicho's ways of doing things are rooted in person- and site-specific know-how. According to Professor Kazuo Koike's interpretation of shop floor practices in postwar Japan, the knowledge that workers share is more implicit than explicit and, as a consequence, know-how held in common through job rotation, on-the-job training, and QC activities may not be readily transferable in terms of any objective description and analysis of the work.[9]

As one of Toshiba's leading factories for the manufacture of cutting-edge, high-tech electronic goods, much of what happens at Yanagicho is not codified and formalized in written documents. The person- and site-specific nature of know-how is a consequence of extremely low turnover in personnel and extremely high levels of functional interdependence in product design, organizational structure, and manufacturing engineering. Knowledge Works embody, but do not necessarily encode, the experience and efforts of its employees.

Except for Accounting and General Affairs Department personnel, virtually all regular employees at Yanagicho are lifetime Yanagicho employees. Although university graduates with technical degrees are hired centrally by Toshiba's Personnel Department, in fact they are likely to spend their entire operational careers in a single factory and product department. Employees without university training are hired directly by Yanagicho and understandably remain there for their work lives, except in those rare cases where a geographical shift in production requires them to move.

The combination of long careers in particular factories and specialized experience with particular product lines makes for highly personalized know-how and experience. Since corporate, nontechnical managers have very different career paths, one where job rotation between different functions and facilities are the norm, it is not surprising that corporate planners underestimate the degree to which site- and relation-specific know-how underpin Knowledge Works.[10] The culmination of such know-how is a factory, like Yanagicho, that works extremely well in its own right but works in a way that is hard to explain, emulate, and transfer abroad. As Mr. M. K. explained, "Certain people know how to do certain things extremely well. As a result, unless everyone cooperates and works well together, the system won't work as a whole."

Screwdriver Factories at Home and Abroad

Considerable discontent has been voiced lately among European Community (EUC) nations against the practice—thought to be particularly Japanese—of setting up so-called screwdriver plants. These are factories that assemble rather than manufacture products. As chapters 2 and 4 have demonstrated, many factories of major firms are clearly that. The division of labor and specialization of function have progressed to the point where suppliers do most of the actual manufacturing, while large companies, like Toshiba, engage primarily in design, development, and final assembly operations.

The degree to which products are manufactured in-house or outsourced is not fixed but depends on a number of interrelated issues, such as unused or underutilized plant capacity, the cost of labor and materials, lead times for delivery, technological capabilities, and managerial resources. Fully two-thirds of Yanagicho's production value in 1986 was contained in the cost of outsourced parts, components, subassemblies, and services. Put another way, Yanagicho itself provides only one-third of the net value added to its products.

Furthermore, typically two-thirds of the goods bought from suppliers were custom manufactured according to specifications provided by Yanagicho. In other words, most of what Yanagicho buys are not off-the-shelf goods, but are made according to Yanagicho's drawings, blueprints, and instructions. As a result, Yanagicho and its suppliers are interdependent, with detailed design data and manufacturing specifications flowing in one direction and high-quality, finished goods arriving from the other.

In the case of PPC manufacture, an extreme example of supplier dependency, Yanagicho provides only one-tenth of net output or value added (for 1987).[11] The greater dependency on suppliers in the case of PPCs springs from a number of sources, including the complexity, precision, and large number of parts that go into PPC production. Also the general tendency toward shorter product life-cycles for PPCs forces Yanagicho to concentrate more on product design and process engineering than on manufacturing. The proliferation of PPC models in the marketplace balloons in-house managerial and manufacturing resource requirements to unacceptable levels. For these reasons, Yanagicho's value-added contribution to PPC manufacture is below average for the factory as a whole.

Rather than attempt to reverse its dependency on outside suppliers for PPC production, Yanagicho is deepening that interdependency. It concentrates its own resources in high-value-added, strategic activities, like hardware and software design, printed circuit board design and manufacture, the production of key components, and the management of outside suppliers. The last is done primarily through the Purchasing Department and it involves selective training of suppliers in technical, organizational, and managerial skills and methods. Generally, it is far more efficient to lower production costs on the supplier side than for the factory to struggle to do so internally.

In spite of the factory's growing dependency on suppliers, Yanagicho reserves for itself certain critical manufacturing tasks, like improving existing products, designing new ones, and enhancing their manufacturability. Primarily but not solely for these reasons, Yanagicho continues to retain final assembly operations. If the entire process of PPC production was outsourced, Yanagicho may no longer understand how products could be improved and where the resources for doing could be found.

Given the extreme division of labor in PPC production at home, it is not surprising that overseas sites for the manufacture of PPC are primarily assembly operations. In this regard, PPC production is rather more complicated than the manufacture of commoditylike products with fewer and less exacting parts, such as televisions, refrigerators, and radio cassette players. So, in the case of PPC manufacture overseas. Toshiba does not have the overseas capability to make most of the parts, components, and subassemblies that go into its PPCs. Nor should it be expected to. It would be ill-conceived to remove the strategic design, development, and engineering capabilities for PPC manufacture from Japan. Toshiba would quickly become an also-ran in PPC sales at home.

But political pressures are another matter. For failing to keep the local content of their European-produced copiers to a level of at least 40 percent, Toshiba and other Japanese firms, such as Matsushita and Konica, have been charged with dumping. Matsushita, for example, was found to be sourcing all but 1.6 percent of its copier components from Japan in February 1988. Toshiba and Konica were not investigated at that time, and by the following fall and winter both claimed that they had cleared the 40 percent level.[12]

Hence, home country and overseas manufacturing are now moving in opposite directions: at home, the pull toward assembly factories for PPC production seems inevitable, given cost and strategic advantages of outsourcing, while abroad the push is toward raising local content. In other words, the thrust of domestic PPC manufacture has been in the direction of what are pejoratively called "screwdriver plants." Nevertheless, what is strategic at home is politically unacceptable in many parts of the world, lending more weight to the problems of internationalizing Knowledge Works.

Organizational Learning and Knowledge

In general, going overseas or internationalizing operations reduces flexibility because, as this chapter argues, the complexity and versatility of manufacturing systems as they exist in Japan cannot be easily transferred abroad. The nature of factory know-how is not contained in manuals but is found instead in practice and experience. This history is embodied in factory-specific, face-to-face relations, on-the-job training, and in people-based, site-specific knowledge. Complex and sticky knowledge, in turn, is rooted in the principle of organizational learning in which effective, usable learning concentrates and resides in specific work sites, functions, and interactions. Such knowledge cannot be simply transferred elsewhere. This may be true regardless of the amount of effort expended to do so, as some examples may demonstrate.

In August 1988, during a meeting with the group of six Yanagicho employees who would be dispatched to establish PPC production at the Irvine, California, site, I asked the Yanagicho team why they had not contacted the Toshiba employees from the Ome Works on the Western outskirts of Tokyo who were already working at Irvine. Surprisingly, they rejected the proposition. With a hint of incredulity in their voices, they wondered how one could learn something useful from someone who has been working in another (Toshiba) factory.

The pride of Yanagicho employees is legend, and these men, chosen to take Yanagicho's most important product to America, must have felt especially proud of their selection and assignment. But pride was not the reason why they resisted asking Ome factory employees about life in America, especially because Ome workers were perfectly able to advise them about when to put out the trash and where to buy a good used car. Instead the impracticality of querying Ome workers lies in the specificity of factory knowledge. To the degree that the principle of organizational learning captures the essence of Yanagicho, learning rather than pride explains the reticence of Yanagicho employees toward their fellow Ome workers.

Another example. The way of detailing engineering drawings between factories is different. The drawings themselves are similar enough but without explanation of what are assumed to be commonsense terms at Yanagicho, the drawings could not be read successfully in France or the United States. Simply put, it is not adequate to translate factory documentation and rules without the background material that informs them. And these being part of the commonsense world of Yanagicho—Yanagicho's way of doing things—it is difficult for anyone not at Yanagicho to distinguish what is significant and what is not, and thus what should or should not be translated before being transferred overseas. Obviously, no one has any idea about what to do or how to do it at the division or higher levels of the firm.

Another example of Yanagicho-specific knowledge can be found in the *fukuju-gijutsu,* or mixed technologies, that characterize PPC production today. Generally, two types of PPC makers compete in Japan: those that evolved from camera production and those that developed from the manufacture of precision machinery. Toshiba falls into the latter category, and as a consequence, much of the technologies of paper transmission, optics, light measuring sensors, microelectronics, and image processing that would be second nature to a camera-making company are foreign to Toshiba. The interworkings of these technologies have to be perfected through trial-and-error at every stage—design, development, manufacturing, and marketing—a perfect recipe for organizational learning. In a very real sense, therefore, the major task of any Yanagicho team destined for overseas assignment is to disentangle the essentials of Yanagicho's way of working from the country- and company-specific context of the Yanagicho Works. Not a simple matter, by any means.

The result, as can be imagined, was an enormous effort in defining tasks, tools, templates, and techniques that imbue Yanagicho's managerial and manufacturing systems. In one sense, they are succeeding, as evidenced in a 15-minute, English-language video on Toshiba's Total Productivity campaign, which is earmarked for overseas training programs. Though the video has its humorous moments, such as when the correct manner and demeanor for wearing the factory uniform are illustrated by having the camera pan slowly from the head to the toes of a Toshiba worker standing at attention, it successfully explains certain aspects of Toshiba quality management programs. But, in another sense, such efforts rarely succeed because what can be described is not what is embodied in the history and human relations of Yanagicho. Going abroad will necessarily reduce the very qualities that have made Yanagicho successful in the first place.

Cost Considerations

In 1988, a square meter of bare land in the Tokyo suburbs went for anything from $80,000 to $120,000. Although land prices and the costs of living in urban Japan are among the highest in the world, Knowledge Works sited in rural Japan are as nonsensical as mass production factories located in downtown Tokyo. If responsiveness to market and technology conditions is crucial in today's rapidly changing world, and everyone says that it is, then being anywhere but on center stage is a serious disadvantage. So, Knowledge Works are naturally found where technology transfer and rapid economic growth are most acutely felt, and these are the megacity nodes of an increasingly interconnected, high-tech world.

As a result of these "Tokyo Effects," a variety of products that were perfected for high-volume manufacturing at Yanagicho are shipped elsewhere to lower-cost centers for mass production. Outstanding past examples of this process include vacuum tubes, radio receivers, calculators, washing machines, refrigerators, freezers, tape recorders, consumer audio equipment, microwave ovens, and specialty metals production. Knowledge Works act as transmission belts of ideas, people, and products from an information-age, urban core to a more traditional, less frantic, rural Japan. Today, processes and products less sensitive to change and more easily mass produced are also being shifted offshore to Southeast Asia, North America, and Western Europe.

Reduced flexibility in organizational capabilities, however, means that a relatively limited line of PPC models can be produced abroad. This undermines some of the principal advantages of Knowledge

Works in Japan, such as the economies of scope realized in the manufacture of a full-line of products in a single facility. Economies of scope yield lower per unit costs because of savings in joint product development, design, manufacture, management, and marketing. Going overseas diminishes the flexibility on which the system depends at home. Various local content requirements further reduce flexibility in manufacturing techniques and product-model selection.

To make concrete what the loss in flexibility might mean, consider the following figures on labor cost differentials, based on average wages for 1987–88. The average hourly wage for employees of Toshiba's Fukaya Works, a color picture tube plant, was 1,800 yen per hour ($13.33 per hour at $1=135 yen). Average factory wages in a principal supplier to Fukaya were 1,300 yen per hour ($9.63). Suppliers, then, paid some 25–35 percent lower wages. However, average wages in Matsushita Kotobukiya's rural, color picture tube facility were reputedly 850 yen per hour ($6.30), while in South Korea's Samsung color picture plant the average wage was 230 yen per hour ($1.70).[13] Cost estimates have to be increased by at least 50 percent since 1987–88, and by another 35 percent based on exchange rate movements in 1995 alone.

Ease of manufacturability is another concern. At home, with a large staff of industrial R & D engineers, considerable attention can be given to ease of manufacturability at initial product design and development stages. But Yanagicho must design PPCs for overseas production without upstream resources in France or the United States. Since not everything about how PPCs are made and assembled can be written down, and since the pace, design, and content of work are different overseas, all of a sudden surprisingly simple matters become fairly complex.

Some of the most vexing matters for Yanagicho engineers were the height of the transfer line and the amount of room needed for the wrists and fingers of an average French or America worker in assembling PPCs. Given the spate of discrimination suits filed against Japanese firms overseas, gender differentials, such as average height and weight differences for male and female employees, have to be considered and planned for. As a result of these complexities, Yanagicho designed a PPC for ease of manufacturability by Americans at Irvine. Because it had fewer parts, it was easier to assemble; because it had fewer parts, it was less functional, that is, it did not do everything that a new PPC in Japan might do, such as enlarging and compressing images or printing on both sides of the page. Adjustments to the height of transfer lines, parts count, and functionality reduce overseas flexibility, raise the cost of going overseas, and clearly weaken the competitive position of Toshiba's PPCs in the international marketplace. And to the degree that the minutiae of PPC production have to be clearly and comprehensively thought about and written down, this too will raise overseas production costs.

Other considerations include the cost of hiring, training, and advancing French and American nationals in Toshiba's overseas manufacturing facilities. Given a fairly positive image of working for the Japanese, overseas recruitment has not been that much of a problem. In nearly six years of operations at Toshiba Systems France and the former 3M toner plant in Mitchell, South Dakota, turnover rates have been quite low. Because these are assembly rather than manufacturing operations, the quality of work is reportedly quite acceptable, given a basic and ongoing education in quality control concepts and methods.

But real problems may emerge in several years, and they are likely to appear in two ways. First, Toshiba's systems of management and manufacturing depend heavily on internal promotion and in-company training, and in this Toshiba is representative of major Japanese firms as a whole. Given higher rates of voluntary and involuntary turnover in Western labor markets, can Japanese firms afford to maintain classical in-company training and "lifetime employment" policies? Further, if firms begin to bid competitively for the services of middle- and upper-level managers to fill out and upgrade the quality of personnel in their international operations, how will this affect the attitudes and ambitions of managers who are following more traditional paths of internal advancement based on in-company education, on-the-job training, and functional experience.

Everyone in Japan believes that they can rise within the organization. Even without a university degree, one can become a *gijutsusha* (engineer) in a company like Toshiba. Which is to say that internal status differences have little to do with formal schooling but have a lot to do with in-company education and experience. This, of course, is consonant with what has been said previously about the importance of organizational learning and the embodiment of factory capabilities in history and human relations. Second, the progression from assembly to full-bodied manufacturing overseas will not be an easy one. Because it is hard to justify the decision to internationalize PPC production on purely economic grounds, interunit and interfunctional political hurdles must be cleared before the economics of overseas PPC production can be tackled head on. Without an understanding of and support for a decision to go overseas, the decision will not likely be made.

Organizational Mismatch

Aside from PPCs, LBPs, rotary compressors, IC card products, and meters for measuring and reporting utility usage, all of Yanagicho's products are produced by custom order in small- to medium-size batches. Such products include automatic mail sorting equipment; ATM machines; rail and subway ticket printing, validating, and processing devices; and optical-electric mass storage image processing systems. By and large, demand for these goods is identified first by a divisional sales office, and then the order is increasingly well specified long before a factory becomes aware that a commission has been secured. Loose coordination is acceptable, even advisable, in these cases because labor saving, automated products of this sort do not involve much preplanning for the market, while at the same time, in Toshiba's way of doing things, they are designed and manufactured by the same working group or team in the factory. Also, there is almost always a significant degree of cooperation between customers and a manufacturer in the case of custom-order goods, so that ATM machines, for example, can run only on software supplied or specified by the customers placing an order.

The process of marketing, designing, and manufacturing PPCs, however, is quite different and requires close coordination in all marketing and manufacturing functions while, at the same time, involving a significant degree of specialization at both the divisional and manufacturing levels. In the case of custom-order goods, by contrast, it is the job of the product planning department in the business unit and the factory administration department (*seisan kanribu*) to coordinate design, development, marketing, and sales. Generally they do it very well. However, the necessarily more difficult task of close coordination in the case of PPC manufacture has collapsed functional specialization between the division and factory, and enhanced product-specific specialization in design, production, and planning within Yanagicho's PPC Department.

In short, a matrix conflict between authority/responsibility and knowledge/experience in the case of Toshiba's overseas PPC manufacture has evolved out of the complex organizational linkages that characterize a multifunction, multiproduct, and multifocal manufacturing facility like Yanagicho. In practice, Yanagicho has responsibility for both divisional and operational aspects of PPC manufacture in Japan while, in the case of Toshiba's decision to take the production of this project overseas, Yanagicho was relegated to second-team status without divisional and operational importance.

This raises another difficulty in the process of internationalizing PPC manufacture. Because Yanagicho is a Knowledge Works as opposed to a mass production facility, it is an extremely complex, interdependent, and yet flexible organization. In terms of its overall disposition of personnel, fully 20 percent of the factory employees work in the design function and another 20 percent are split between manufacturing administration (*seisan kanribu*), manufacturing engineering (*seizo gijutsubu*), and product/production technology (*seisan gijutsubu*).

In PPC production at home, there are seven levels of administration from shop floor to department head, and only the last of these, department head or *bucho*, has anything to do with formal education or university-level training. Only 4 or 5 of Yanagicho's 23 *bucho* do not have a university education. However, fully half of the factory's *kacho* or section heads are high school graduates, and practically all of the foremen are high school graduates. Thus, from shop floor workers up through team leaders (*ri-da*), working group heads (*sagyocho*), assistant foremen (*seizocho hosa*), foremen (*seizocho*), and section heads (*kacho*), factory-specific training and experience are critical in developing know-how in PPC design and manufacture.

Neither is Yanagicho 100 percent free of the design and engineering capabilities of its suppliers and, as a consequence, the factory may be severely constrained in its selection of models for overseas manufacture. Further, given the shortening of PPC product life-cycles in Japan and the need to be readying the next generation of product as the current generation is going into production, the opening of overseas manufacturing facilities both reduces the resources that can be devoted to new product development at home and distances overseas operations from the aggregated, localized resources that have made for successful PPC research, design, and production at home.

Development Is not Abstract Art

"Unlike basic research, development is not an abstract thing," according to Mr. Tomonobu Shibamiya, a Toshiba engineer in California. "To develop new products, one needs manufacturing experience with previous products, some design seasoning, contact with suppliers, a sense of how applications differ. These ingredients are home-grown and not pulled off-the-shelf."[14] Home-grown means that products are *locally* designed, *locally* built, and *locally* tested according to *local* standards and sensibilities.

He continued, "*Kaihatsu* (development) refers to technology and materials. These are difficult enough to evaluate at home, but they are nearly impossible to evaluate overseas. Japanese-style development often involves severe time constraints, making prototypes by hand, and rigorous evaluation. How can we reproduce these conditions abroad when only one part of the Japanese team (i.e., manufacturing) goes overseas?"

Shibamiya was referring, of course, to the team of technicians, designers, production/process specialists, suppliers, and sales engineers that work together during new product development in Japan. Integrating functional know-how during product design and development is the key to making products with "integrity."[15] Integrity appears closely connected to the organization, composition, and leadership of product development teams, and hence, to the products they bring to life.[16] Process issues, such as these, come together and are resolved in Knowledge Works.

In the case of a medium-sized PPC, there are about 1,000 assembly parts and steps in the process. Given this complexity, the product does not lend itself easily to computer-aided design (CAD) techniques or to abstract design approaches. Local know-how and experience in both product design and manufacturing process are essential for successful product development. Just as clearly, the design, development, and assembly know-how on which PPC cost and performance depend cannot be easily or quickly transferred overseas.

But the EUC's 40 percent local content rule means that Japanese manufacturers have to raise local content levels quickly by a hurried selection of local vendors. As a result, the quality, performance, and cost of Japanese photocopiers manufactured in the EC suffer by comparison with those manufactured in North America. The absence of local content legislation in North America allows Japanese makers more time to train local assemblers and engineers, to recruit and educate local vendors, to emphasize an *accumulation* of process engineering know-how, and to develop more complex, higher-value-added products.[17] Of course, without local content requirements, Japanese firms may simply import most of the higher-value-added parts and components that they need.

Since PPC manufacture is characterized by mid-sized lots of extremely high quality, say, 1,000 units per month of a certain size and performance profile, it is extremely difficult to recruit local vendors who can supply such high-quality parts and components in relatively limited numbers. The investment needed to meet quality, performance, and cost requirements is considerable, and there is not enough margin in the business to overcome the reluctance of vendors in being so decidedly committed to one assembler.

Design, Teamwork, and Problem-Solving

Mr. Shibamiya became the plant manager at Irvine after spending two years at another overseas plant (Mitchell, South Dakota). He confesses, "We are not so global in our operations. Our molds and dies still come from Japan, and this is likely to continue as long as the production runs overseas are, at best, one-third as long of those in Japan." In other words, local production runs are not large enough to make it economical to source molds and dies locally. In Japan, molds and dies for a particular PPC model might be used for 12–18 months with a popular PPC model selling around 4,000–5,000 units per month. Irvine was running at 1,000 units a month for all models at the time.

Molds and dies come from Japan for a very simple reason: there are no PPC design engineers in Irvine. Their absence is not an act of prejudice but one of tactics. There are not nearly enough young, talented design engineers in Japan and, of these, relatively few want to go overseas. In fact, few of them do. So, it makes good sense to concentrate engineering talent in Japan where they can be more effectively trained and employed. But, as a result, it is nearly impossible to achieve cost reductions overseas through better engineering, better design, and better manufacturing practice. Indeed, it is difficult enough to stabilize production and to achieve consistency and reliability in manufacturing operations.

Besides an absence of design engineers, Mr. Shibamiya bemoans the quality of shop floor workers. Recalling his previous assignment in Toshiba's Mitchell, South Dakota, plant that makes PPC toner, he reflects, "Sometimes, nearly insignificant skills can make a difference. Last year, we had a toner quality problem . . . in the production of toner. In theory, toner is toner but, in fact, lots of little problems creep up, like temperature and humidity variations that can affect quality. At the design stage, engineers work hard to minimize such problems by making appropriate specifications in their drawings, blueprints, and testing procedures but, obviously, they cannot think of everything. That's where experienced and motivated workers make a difference. Without teamwork between engineers and workers, we could not have solved the toner problem. Fortunately, we did so but, in fact, toner is not such a complicated product. It doesn't demand high levels of coordination in design, engineering, and production functions, like PPCs do.

"Without teamwork, in fact, most manufacturing problems cannot be solved. In Japan, teamwork is part of manufacturing practice. But here, I don't know how foreign engineers are supposed to work, and I don't know if they can work effectively with shop floor workers and suppliers."

Shibamiya's thoughts about the difficulties of localizing manufacturing functions were echoed by Jun Kawasaki. He added, "Originally, we had a three-year plan to produce two PPC models here. But we added 2 additional models on a OEM [original equipment manufacturer] contract basis, and this threw off our schedule. By adding models, the volumes originally predicted dropped, and not surprisingly prices went up. Also, assembly cycle times shot up because workers had to learn twice as many assembly and testing routines. So, all our calculations about quality, time to market, and cost were thrown out the window."

Irvine had only 15 locals assembling PPCs in 1990. (A short PPC line at the parent factory would have about 40 assemblers.) The small number reflects difficulties in recruiting local workers as well as in deploying them among the different lines at Irvine. PPC assembly competes with the assembly of digital phone exchanges, facsimiles, x-ray machines, and laptop computers for the available labor supply. Unless PPC model volumes can be increased notably and workers allowed to specialize in the assembly of no more than two of the four PPC models, it is unlikely that 15 locals will learn well four different assembly routines. In fact, this is asking more of them than Yanagicho workers. Given the complexity of these multiple tasks, Irvine is operating at 60–70 percent of standard operating times in Japan, and struggling.

Mr. Susumu Domon, a manufacturing engineer, chips in. "The biggest problem is lack of stability in the assembly process. Without enough volume and reliability in assembly operations, workers are not able to make independent, experience-based judgments about problems that arise on the job. For example, they're not able to judge when an existing problem stems from a locally supplied part, the assembly process, or the manufacturability of the product itself." In effect, production volumes are too low to generate sufficient learning for assemblers to discriminate and diagnose problems. Problem solving proves impossible.

"Worker thinking is digitized," he continues. "Problems are either black or white for them. Yet for us, problems are neither black nor white but shades of grey. Shades of grey call for analogue thinking—reasoning about how this problem is more or less like something else you've experienced before. They don't have enough experience for analogue thinking."

Domon worries out loud that the choice at Irvine is between better qualified people or more automation in the assembly process. But, given the low volumes involved and PPC model variety, robotization and automation are simply out of the question. He concludes, "In the end, we have to create a factory culture that will support, motivate, and enliven the workforce. I'm afraid that's the only answer."

Factory manager Shibamiya jumps back in. "Back home, in spite of many layers and levels in the corporate organization, there is one person responsible for each product. [Each product is "managed" by a specific business unit [BU] product manager.] When we have a problem, we can go to this guy and he gives us a solution. But that's an impossibility overseas . . . at least, for now." Shibamiya shifts gears, "Last year, we lost twenty percent of our personnel in turnover, and that's not counting interdepartmental transfers which add up to another twenty percent. How can manufacturing practice get better when forty percent of your personnel disappear in a year?" The Sunday edition of the *Los Angeles Times*, to confirm Shibamiya's point, carries page after page of classified announcements for assembly and production workers.

Bottlenecks and Blueprints

Domon is in charge of shop floor operations for PPC assembly. He recalls his "culture shock" when he arrived at Irvine. "It takes about three months for a worker in Japan to get up to speed on the line. I thought that after a year here, we could be at 80 percent of the standard times at home. But, after a year, we've only reached 50 percent of the Japanese standard."

He continued, "Assemblers here forget what they learn from one day to the next. Partly, they don't have the skill and ambition of workers back home but partly our volumes are too low for them to get sufficient product-specific experience to become competent assemblers. Overall, across the four models we assemble, we're at 30 percent of the model volume in Japan. That's just too low.

"Too low stretches a worker's experience too thin. In order to get 'focus' on a particular product, you need volumes on the order of five thousand units per month. We're nowhere near those levels. In Japan when demand peaks, we can achieve those levels by shifting people within the plant, if necessary. But shifting people is not feasible here. Good lateral communications and information exchange are not so common."

Performance measures are so different that Domon relates how the Irvine yardstick is the number of hours required to assemble one unit (hours/unit) whereas at Yanagicho, it is the number of minutes to assemble one unit (minutes/unit). Part of this surprising difference springs from the six nationalities represented among 15 different PPC assemblers. (They don't have a common language of social experience, training, work experience, or communication.) Also, turnover on the PPC line was running at 3 percent a month in the early 1990s. Workers were young, culturally heterogeneous, averaging 27 or 28 years of age, and with lots of opportunities to move locally for work.

In Japan, by contrast, the average age of PPC assemblers at Yanagicho is 42 years of age. Besides being culturally homogeneous (in the national sense), they are correspondingly more experienced, more settled in their personal lives, and more anxious to be successful on the job. In order to encourage low turnover and low absenteeism at Irvine, 100 percent attendance for an entire assembly team raises everyone's wages on the team by $0.50 an hour. Not surprisingly, this inducement is unneeded in Japan.

Obviously, high turnover and low volume make it hard for workers to gain experience of PPC assembly. It also makes it hard to teach better practice. "It's a vicious circle," says Domon. "Volumes are low and turnover is high. As a result, workers lack personal experience on the line and they lack experience working with each other. The tentativeness of the situation makes it difficult to teach anyone anything. Suggestions about how to do something better are taken as personal faults and public criticisms.

"PPC assemblers in Japan, accustomed to working together, will say, 'Hey, you missed a screw here or there's something wrong with the alignment of the part you installed.' By common agreement, it is believed that production steps are linked and integrated. *Mae kotei* [upstream work] and *ato kotei* [downstream work] are inextricably linked. An individual worker's responsibility extends all the way up and down the line, including *mae kotei* and *ato kotei*. In such an environment, telling your neighbor that something is wrong is not a criticism. It is an act of mutual responsibility, and it benefits everybody, since upstream and downstream work are really one and the same thing. Understandably, that attitude is lost overseas where labor conditions, production volume, and employment persistency vary considerably."

Ideally, the low volumes at Irvine could be circumvented by localizing production for one or two PPC models there and by assigning responsibility for the other two models elsewhere. Reducing model variety will increase production volumes for the remaining models. But several factors frustrate this solution. PPC sales worldwide are increasing at only 3 percent a year. This is not enough to justify the localization of a full staff of design, development, and manufacturing engineers at several different locations worldwide, as the transnational model of regional organization and coordination would suggest.

Also, local content requirements differ by region. This complicates a worldwide division of labor because the supply ratios of locally sourced, high- and low-value-added parts and subassemblies are not consistent across regions. Economies of scale that might be achieved by localizing the manufacture of a key component, like printed circuit boards (PCBs) in one site, are frustrated by a need to localize production of the same component at several different sites. Local content rules play havoc with global sourcing and manufacturing strategies.

Mono O versus *Hito O Genchika:* Localizing Products or People?

Mr. Domon says that localizing products is relatively easy but that localizing people is pretty hard. What he means is that products can be taken out of Japan without too much difficulty, but it is much more difficult to take people out of the products. Or, in other words, people make products. People make products in the obvious way, that is, with their hands, but people also make products in less obvious ways, that is with their minds, hearts, and souls.

While it should be possible to make the same products in a different organizational environment, Irvine has not figured out how to do so. And, for the moment, changing the organizational circumstances is not being pursued as a solution because the attributes and attitudes of the local workforce are far more salient to Japanese managers. Obviously, some combination of the two—changing organizational arrangements as well as worker attributes and attitudes— may improve factory performance.

To the extent that products are "people-dependent," it will be difficult to internationalize their manufacture with anything near the same levels of industrial quality, ingenuity, and performance that are realized in Japan. The problem, according to Domon is that "local staff are passive. They will accept instructions but they won't think on their own about what it will take to make things better. What is missing is *yarigai,* or an active participation in the process of work. It's a chicken and egg problem. If we could get people into the products, the products would improve. If the products improve, more investment will come from Japan. More investment from Japan will let us localize more products and more of the product design and production process. *Yarigai* [desire to do the job] will no longer be a problem."

"Even though a few people at the factory back home understand the problems, at the level of the business unit (BU), salesmen and nontechnical managers do not. They simply can't understand why we're taking so long to get on track. But time-on-task is not high. Assemblers' attention wanders. They talk to each other, and they lose concentration. The PPC line is supposed to be a no-talking line but over in the East Wing where they assemble facsimiles, they talk all the time. It's impossible to enforce a no-talking rule here."

Domon continues, "All of the reworking [to correct assembly defects] is because of simple assembly mistakes. Because the assemblers talk to each other, they don't develop rhythm and concentration. Overall productivity levels are low. In Japan, a supervisor could say 'that was dumb' [nani yattenda!] and that would solve the problem. And the employee would apologize. But here, you can't call someone a fool or say 'that was dumb.' The only option is give a gentle warning the first time, a more insistent warning the second time, and then a dismissal notice."

After a moment's hesitation, Domon whispers, "There's only one assembler out of fifteen who can do more than one job. [In Japan, it's common to learn all the jobs within a work team on the line.] Even so, here, they all want to change their jobs . . . without mastering their current jobs." He stops talking for a moment. "Several years ago at the main plant in Japan," he continues, "a 'madam-line' [of mature, part-time female workers] was formed. They performed better in quality and productivity than the main line. So, from a Japanese point of view, PPC assembly is not so difficult. But, at present, assembly productivity is at least four times higher in Japan than here. So unless volume increases dramatically or PPC assembly becomes much simpler, we can't expect much improvement.

"The great thing about Japan," he concludes, "is *kyoryoku kojo,* or the factory that works together. Everyone tries to understand what was the design engineer's intention when looking at the blueprints. And if after making some effort they still don't understand, they call him over and have him explain what he wanted. At Irvine, without a perfect design, nobody knows what to do. And since nobody wants to ask questions, show that they don't know, and get involved personally in their work, a lot of poor-quality workmanship and products result."

At the time of these interviews, Toshiba's Irvine plant had been in operation for nearly three years. While everyone seemed to agree that hiring some local design engineers would help solve problems, no one expected that to happen very soon. Non-Japanese design engineers will not have any experience in designing PPCs, a product virtually monopolized by Japanese makers. While non-Japanese engineers could be hired locally and sent to Japan for training, it would take about three years in Japan before a non-Japanese engineer could become a competent PPC design engineer.

Making PPCs: Global or Local Products?

Knowledge Works are becoming more common and strategic in Japan while, at the same time, their site- and relation-specific capabilities cannot be easily transferred overseas. They function well because of high levels of on-site training, investment in human resources, low levels of employee turnover, the active involvement of employees in work design and content, union-management agreement, and site-specific teamwork in engineering activities. In brief, Knowledge Works function well because of a factory culture rich in learning, meaning, content, and context.

If site-specific social and organizational factors are key features of manufacturing success, they are even more pronounced in factories where mental more than manual work predominates. So, there is not a "Hitachi way" or a "Sony way" but a number of different, factory-specific ways within Hitachi and Sony. Hitachi, of course, may decide to export "Factory A's way" but, for now, matters of subunit specificity have been overshadowed by corporatewide issues of "Hitachi's way." Whether or not this is the correct approach to global strategy is another question.

Industry matters.[18] Industry characteristics may make it more difficult for Japanese TNCs in electronics to successfully transfer their upstream activities and facilities overseas as compared with motor vehicle makers, for example. Industry differences are important as well because most of the writing about Japanese transplants to date has focused on automobiles or consumer electronics rather than industrial electronics and information/communications technologies.[19] In today's overwrought search for global firms and products, it is easy to forget company-specific and site-specific characteristics that are the real sources of industry performance and competitive advantage. Indeed, designing PPCs, even more than making them, requires complex organizational, managerial, and systems integration capabilities—precisely the kind of things that are hard to transfer overseas, and exactly the things that are easily forgotten in discussing global strategy.

In Japan, manufacturing facilities are increasingly differentiated according to upstream and downstream production tasks. The process of differentiation is both physical, in terms of investment in plant and equipment, as well as social and organizational, in terms of how people of recruited, socialized, trained, and managed. Given the absence of these factors at Irvine as well as the low production volumes, low levels of learning, and low complexity of the assembly tasks, even downstream functions, like reliably assembling PPCs, seem questionable.

In theory, a photocopier is a global product or close to one. It requires a moderate level of adaptation to local conditions, as reflected in different electric current requirements, different sizes and grades of sheet paper, different temperature and humidity conditions that affect toner and electrostatic imaging. Or, in short, a moderately differentiated approach to worldwide marketing. At the same time, global sales of PPCs require a fairly high level of organizational and managerial coordination between overseas subsidiaries and the parent company, so that a moderate-to-high need for corporate integration is evident. High or relatively high levels of differentiation and integration suggest a product strategy that is generally considered a global one.[20]

The difficulties of globalizing PPC manufacture, design, and development are manifest, however. Are these difficulties unique to the product or the production system? When the long list of Japanese success in digital facsimiles, engineering work stations, laser beam printers, private branch exchanges (PBXs), optical-electric storage systems, and other high-tech, communication and information technology wonders are considered, it seems as if the PPC story is not an isolated one. These products may be more or less global, but their production systems are not. They demand a high order of corporate capability as well as factory-specific design and development know-how. These are ultimately embedded in workplace cultures, as negotiated and expressed at sites like Yanagicho.

Finally, variations in how product lines are organized and managed allow a range of options with respect to how easily parts of the value chain are separated off and transferred abroad. Champion Line functions are tightly integrated on-site, so this complicates a disaggregation of design, development, and production functions. In the Balanced Line model, on the other hand, different functions and various steps may be more easily hived off to affiliates and suppliers, at home and overseas.

Internationalizing Yanagicho

This chapter, based largely on fieldwork interviews in Japan, France, and the United States, describes a number of interunit and interfunctional problems encountered when Toshiba decided to internationalize the manufacture of plain paper copiers. For the most part problems arose in three ways. First, corporate level planners were not fully aware of the differentiation of manufacturing functions within Japan, including distinctions surrounding product design and development processes as well as the nature of supplier interdependency. Second, because of this, corporate level planners were not able to provide for the effective transfer of key operational features and capabilities that underpin domestic manufacturing. Third, as a result, the overall success and the relative speed and effectiveness of international organization and technology transfer overseas were hampered.

Toshiba is currently producing PPCs in three countries outside Japan and supporting PPC production by foreign firms in two more. What began as a technology licensing agreement in China has been twice upgraded to a majority-owned firm that is challenging TSF's Normandy plant for the honor of being Toshiba's leading overseas producer of PPC. Yet Toshiba's French plant is doing well too. When a second PPC model was added to the production line in France in 1989, production volumes dropped precipitously. But with diligence and effort, volumes picked up and were back to normal within three months. Today, Toshiba Systems France produces four models of photocopiers as well as toner at its new facility in Neuville des Dieppe. And in Mitchell, South Dakota, Toshiba's PPC toner plant has been generating record profits and moving to diversify its product line by introducing the manufacture of electronic instruments and controllers.

In this way, some foreign operations are gaining design and development experience independent of Yanagicho even as they have been loyally supported by Yanagicho in that process. However, at Irvine, the PPC line was closed down in 1993, after five years of effort to transfer and implant PPC manufacture in California. The problems of assembling PPCs, even a limited line of PPCs, at current yen/dollar exchange rates, were overwhelming.

The most important problems were organizational and informational: the differences between Yanagicho, a multifunction, multiproduct, virtuoso plant at home, and Irvine, an overseas plant with limited functional capabilities making a wide range of products in relatively low volumes. Also, the sources of organizational knowledge were many—engineers and products from five different factories in Japan were brought together at Irvine—and it was practically impossible to create a single set of norms and standards with respect to how things should be done. Also, PPC production volumes were low, turnover rates high, on-site learning and skilling were minimal.

But PPC themselves are changing. They are becoming more network and computer based. As these changes unfold, overseas research and engineering skills will become increasingly important and in all likelihood they will be incorporated in the design, development, and manufacture of PPCs overseas. The trend seems inevitable even if it is happening more slowly in the case of photocopiers than in the cases of motor vehicles, and consumer electronics.

Nevertheless, the story of how PPC production was internationalized is a case study in a classic yet chronic conflict between autonomy and control in large, bureaucratic organizations. The face-off appeared in an effort to initiate PPC manufacture overseas, the operational aspects of which were delegated to the factory (Yanagicho) while the strategic push came from the Office Equipment Section of the Information and Telecommunications Division. Better coordination between strategic and operational goals might have been expected, given Toshiba's ranking among Japan's better-managed industrial firms.

The organizational complexities of Knowledge Works, which is to say the complexities of multiple, overlapping department and divisional jurisdictions in a multifunction, multifocal factory, have resulted in some loss of corporate control and oversight. A more centralized organization that cleanly demarcates product and market boundaries might be more appropriate for overseas operations.[21] Yet, because Knowledge Works are sites of product and process innovation that depend on the quality of human relations and the effectiveness with which research, design, development, and manufacturing functions are integrated, it is hard to imagine how such embedded and embodied systems may be easily desegregated and shipped overseas. In short, transfer of integrated, highly specific, and personalized manufacturing systems at full strength and functionality is problem-filled.

Industrial firms without Toshiba's highly interdependent manufacturing systems, such as process industry firms, may not face similar barriers to effective transfer of their systems overseas. Dainippon Ink & Chemicals, after their acquisitions of Reichhold Chemicals and Sun Chemicals, is establishing a "four head-office system," with units in Tokyo, the United States, West Germany, and Singapore, all performing head office functions. NSK, one of Japan and the world's largest ball bearing manufacturers, is also moving to a regional headquarters model with Japan, Asia, Europe and North America as the four poles of its global network. Apparently Dainippon and NSK do not anticipate any insurmountable technical or organizational problems in going international.

But what may work for Dainippon Ink & Chemicals or NSK is unlikely to work for Toshiba, given the differing nature of their basic production systems and organizational structure.[22] With a product as complex as a modern, multifunction PPCs, the process of taking production overseas is indeed problematic. At home, PPC manufacture depends on high levels of interunit and interfunctional integration and on site-specific knowledge, skill, and experience. Such resources, patterns, and experiences are particularistic to the manufacturing organization in question.

In sum, Japanese FDI in manufacturing will continue to grow, and the upstream portion of it will increase disproportionately. Such investments will cluster first in products that are relatively simple in a technical sense and that are not too difficult to make and assemble overseas. Also, it is likely that products manufactured overseas will have relatively long product and development life-cycles, say, 30–45 months, and that they will not be too different in product/ process engineering requirements from one product generation to the next.

The degree to which suppliers at home are important in the design, development, and manufacture of products suggests the degree to which a similar division of labor may be contemplated overseas. In other words, products that have a high supplier content at home may, with proper investment and effort, become products with high supplier content overseas. However, as the Champion Line and Balanced Line models suggest, supplier participation may occur in any number of ways and for any number of reasons. It will be difficult to replicate that range of choices overseas. Toshiba needs to remember how Knowledge Works developed in the first place before they set out to transfer and reinvent them overseas.

Hence, PPCs and follow-on products, such as PPCs with networking capabilities and digital, double-sided, color imaging PPCs, are not likely candidates for overseas manufacture. Ironically, therefore, as photocopiers become more of a global product, the high-end of the line are still likely to be designed and developed at home. The site-specific, organizational, and technical requirements of such products are not easily transferable or translatable to non-Japanese work environments. PPCs are global products in a global age but with a decidedly local manufacturing origin and character. So, the "hollowing-out" (*kudouka*) phenomenon is likely to displace workers from relatively low-skilled, production-focused jobs in Japan but it is less likely to affect relatively higher skilled research, design, development, and engineering jobs.

SEVEN

Learning Strategies and Learning Factories

Making many different things with economy, speed, and integrity is Toshiba's manufacturing strategy. To enact this strategy in an increasingly competitive world, Toshiba has advanced fundamentally new forms of manufacturing organization during the past quarter-century, an especially turbulent era for high-technology industries. Knowledge Works, the most outstanding of these forms, fuses best practices from shop floor operations, production engineering, organizational campaigning, and technology management with industrial R&D.

Knowledge Works give Toshiba the means to respond quickly and well to technical progress, exchange rate fluctuations, consumer whimsy, and incessant market movements. Stormy competition, in other words, has prompted Toshiba to equip its leading factories with enough resources to stay ahead of rivals and a strategy to do just that. Knowledge Works do so by mobilizing, motivating, and moving ahead the men and women who discover and apply the intellectual capital that translates into industrial success. Their creativity and efforts culminate in something like an expert system with respect to certain technologies, design and development skills, manufacturing and assembly processes. What Toshiba's Yanagicho Works makes well, such as photocopiers, laser beam printers, and IC cards, is not what other Toshiba plants make well. Yanagicho's capabilities and resources are site- and relation-specific.

Knowledge Works are categorically different from mass production factories that are typically labor and not management intensive, remote from megacity markets, devoted to product/process standardization and the interchangeability of parts and personnel. By contrast, Knowledge Works minimize the time and cost of bringing new and better products to market, maximize product/process innovations, and integrate people-dependent knowledge in management-intensive factories. Embodied intellectual capital is the distinguishing feature of Knowledge Works, and Toshiba's strategy is to express and exploit that capital as promptly and thoroughly as possible.

Knowledge Works Costs

The merits of this model depend entirely on the value of the knowledge generated, accumulated, and expressed on-site. Since the transformation of information into knowledge is costly, its value should equal or hopefully exceed transformation costs. Siting factories in teeming cities, staffing them with lots of technicians, engineers, and scientists, and managing them as multifocal, multiproduct, and full-function facilities is expensive, often extremely so. Transformation and production costs are necessarily high.

Since the organizational knowledge that is expressed on-site is sticky in generation, socially embedded, and imperfectly shared, it is hard to capitalize on and extremely difficult to replicate; witness the trials, tribulations, and occasional triumphs of Yanagicho's internationalization efforts. Organizational knowledge, moreover, is of many kinds, some more amenable to exploitation and transfer than others. For instance, there are four preproduction, transformation costs associated with knowledge: discovery, encoding, transfer, and application. These are irreversible costs. There are also recovery and reapplication costs, which, unlike the previous four, are variable costs. That is, recovery and reapplication costs, have little to do with the time and money invested in discovery, encoding, transfer, and application. Finally, there are significant costs associated with putting knowledge into production or practice.

With respect to Yanagicho's development of plain paper copiers (PPCs), for example, there are many areas of applied research that affect discovery and encoding costs. PPCs incorporate numerous technologies, such as paper handling, electronic controls, image processing, electromagnetism, ink printing, optical character recognition, optics, application software, systems software, and precision machining, to name just the most obvious. There are always new and improved ways that these may be applied to PPC assembly and manufacture; moreover, technologies interact, and interaction affects PPC design, development, and manufacture.

Encoding costs arise when ideas that have already shown some utility or value (at the discovery level of activity) are made more concrete and available. Obviously, there are many ways that encoding may happen; everything from sitting down and talking about new ideas with colleagues to discovery filings, patent applications, and journal publications. Encoding costs vary tremendously and if a poor job is done at this stage, additional costs may be incurred farther down the value chain if and when organizational practices have to be revised and reworked.

Transfer costs have to do with moving concepts, prototypes, engineering samples, or any other intermediate form of knowledge downstream toward production. These costs vary according to the familiarity of designers and engineers with manufacturing issues but, generally, it is safe to say that costs increase markedly when knowledge is moved from a stage of exploration to one of exploitation. Presumably, some ways of transferring knowledge minimize such costs. Knowledge Works embody one of Toshiba's strategies for doing so by grouping together the processes, products, and people that fit together. Fit is measured by how well transaction and production costs are minimized, cross-product and -function learning maximized, and new and improved concepts are incorporated into design and development activities.

Application costs refer to the very real costs of taking even a proven concept, such as an engineering sample, into production. The SuperSmart card's development story provides graphic evidence of these costs as well as the many alternative development pathways that abound. While pilot production may identify some of the problems in applying organizational knowledge in full-scale production, not all volume- and variety-related manufacturing problems appear during pilot production. Because of Toshiba's "market-in, not product-out" philosophy, all product and process changes cannot be anticipated in the course of development. Market feedback is crucial to product refinement and enhancement. Finally, the way that change orders from marketing or engineering are fed back into and modify the store of organizational knowledge affects application costs.

Recovery costs are essentially reverse-engineering costs. Once a concept has moved downstream into production, say how the latest optical character recognition technology is applied in PPCs, what are the costs involved in recovering the technology and reusing it, say in designing, developing, or making laser beam printers (LBPs)? Recovery costs are directly related to how application, transfer, and encoding costs were originally incurred. If some of those costs were incurred in the interest of enhancing the flexibility and accessibility of knowledge-recovery activities in the future, then recovery costs may be lower. Also, hiring retired engineers as consultants to existing projects will likely lower recovery costs.

Reapplication costs are the costs of applying organizational knowledge in a new set of circumstances. Reliably using optical character recognition technologies developed for PPC in LBP without a lot of new experimentation and adaptation is not at all obvious. Thus, the reusability of knowledge, especially downstream manufacturing knowledge with respect to upstream design and development activities, is a core strategy issue. The more reusable organizational knowledge—up and down the value chain—the greater the fungibility of core skills, capabilities, and resources, and thus the wider the scope of potential business applications based on an existing knowledge base.

Finally, production costs vary widely depending on scale and scope effects. Less obviously, learning based on scale and scope effects also varies widely because the array of learning routines and practices will profoundly influence long-term production costs. Here, people are all-important. How experienced, well trained, motivated, and managed are they, especially with respect to ordinary, everyday learning opportunities? Also, the degree to which employees engage positively in TQM and TP practices makes a lot of difference. Besides straightforward learning effects, product/process redesign learning, cross-product design and development learning, learning from cycle time reductions, cost-down and productivity efforts, and other design and development-related activities impact production costs.

In other words, the costs of acquiring, storing, applying, and reusing organizational knowledge are many and significant. It is important to note that such costs are being discussed in circumstances where all the steps and stages of knowledge discovery and application occur in physical and organizational proximity: in Knowledge Works. Obviously, the greater the spatial and organization separation between steps and stages, the greater the costs will be in acquiring, accumulating, transferring, and using organizational knowledge. Knowledge Works are designed to minimize such costs, although discovery, transformation, and production costs are inevitable. Thus, the structures and strategies of Knowledge Works delimit the ways in which organizational knowledge may be created and exploited for new and existing applications. Together, they define the costs as well as the limits of effectiveness, although they do not at all determine the efficiency of Knowledge Works processes.

Knowledge Works Benefits

The Knowledge Works strategy is to amortize the costs of acquiring and applying organizational knowledge by creating and managing multiple product/process domains where the value of what is created outstrips transaction and production costs and where innovation and renewal can arise as circumstances require. So far, it seems to be working. Knowledge Works cut energy and imported material costs during the 1970s, trimmed excess models as the yen strengthened and some niche products lost appeal in the 1980s, fused technologies and launched innovative, new products in the 1990s, such as laptop engineering workstations and full-color, digital fax/photo-copiers.

Other than minimizing costs and innovating new and refined products and processes, speed is another advantage of Knowledge Works. Speed is a measure of the time required to deliver reliable, cost-competitive, high-value-added products to market. Generating more value than costs under circumstances of extreme time-based competition, especially with older workers, plants, and equipment, is absolutely strategic because the all-important question is what should be made as opposed to what might be made. Market pull, not technology push is the answer.

Speed enables Toshiba to retain its competitive edge in time-based competition in spite of the growing strength of the yen and the challenge of newly industrializing countries such as South Korea, Taiwan, and Thailand. Speed is really a summary measure for a host of other Knowledge Works advantages, such as well-trained and -motivated employees, well-organized and -managed supplier networks, accumulated and thoughtful manufacturing experience, process automation and systems integration skills, an egalitarian bias on the line, social goodwill, and a strong local culture. Such advantages, singly and in concert, are not found in low-wage factories characterized by high turnover and low measures of social overhead investment, employee training, organizational learning, and functional and systems integration.[1]

The Knowledge Works model suggests that factories may be a generic, indeed a strategic, way of organizing in volatile, high-technology industries where focus, flexibility, and time to market are critical. In such industries, all manner of resources and capabilities must be localized and integrated in a timely and cost-effective way. While there may be a number of organizational designs for doing this, none other than the Knowledge Works model have been described in detail. Knowledge Works offer a compelling model of how to create and manage organizational knowledge in an age when intellectual capital is our most important asset.

The Role of Innovation

In many large firms in North America and Western Europe, the need to innovate is often associated with merger and acquisition activities. That is, by buying other firms and their products—usually smaller and more entrepreneurial businesses—larger firms find the new ideas, products, and people that they need to sustain their operations. In Japan, by contrast, unfriendly takeovers and venture capital infusions by unrelated firms are rare. Companies must plan for, manage, and sustain internal processes of innovation on their own.

In some crucial ways, intrapreneuring within firms goes beyond the entrepreneurial efforts of venture capital firms. Venture capitalists face cost and time constraints in transforming information into knowledge and knowledge into action. They need to discover what other firms are doing, an inherently opaque undertaking, decide what if anything to do in response, and then act on it. And, of course, those targeted for acquisition or takeover may object, resist, and finally acquiesce, all of which are likely to raise costs even more.

By their very nature, Knowledge Works are intrapreneurial, since they seek to balance conflicting needs for focus, flexibility, innovation, and speed in their manufacturing operations. Such intrapreneuring, by its very nature, is both costly and risky. Each new generation of products and manufacturing processes is likely to be more costly than previous ones as new, capital-intensive product/process technologies are tested and brought on stream. It is not at all obvious what returns will be earned by ever greater investments in technology or in skilled and experienced workers.

Risks and rewards can only be calculated at the end of a full cycle of investment beginning with product design and culminating in market sales. And because product enhancements and refinements from one generation to the next have strong effects on market share and profitability, several cycles of investment and revenue generation may be required before a product's profit potential and spillover effects can be fairly assessed. Nonetheless, capitalizing on existing knowledge is much less costly and risky than trying to promote innovation by taking over another firm's assets and people.

Design & Build: Full Life-Cycle Accounting

In most accounting calculations, the costs of bringing products up to the point of stable and steady manufacture are missing. R & D, design, development, and prototyping costs are typically allocated elsewhere, not charged against the factories where products are typically made. Hence, preproduction R & D and design costs are not normally part of the manufacturing cost equation. But at Yanagi-cho, it is impossible to allocate product development and refinement costs "elsewhere," because sizeable yet indeterminate amounts of such costs are related to being multifunction, multiproduct, multifocal manufacturing sites. In essence, R & D costs are factory operating costs.

Moreover, R & D costs are likely to grow because technical progress requires ever greater investment in people, plant, equipment, technologies, and processes. But, even when all the costs of product design, development, and enhancement are fully and fairly allocated, Knowledge Works more than pull their own weight. The risk reduction capacity of the Knowledge Works model is found in its abilities to cross-subsidize R & D activities and engineering investments across a number of product lines and, simultaneously, to create value up and down the multifunction product design, development, and manufacturing chain. Benefiting broadly from information sharing and learning is the norm, across and within a Knowledge Works' many product lines.

Thus, products of high- and low-manufacturing volume as well as high- and low-manufacturing complexity are accommodated within the model. And, products that demand highly specific code-velopment activities with suppliers, such as photocopiers and IC cards, may be likewise accommodated. Even products that require high levels of on-site product design and process engineering reliability, such as ATMs and automatic mail sorting devices, are accommodated. In short, full life-cycle, site- and relation-specific costs, including research, design, development, and maintenance costs, may be accommodated and allocated in the Knowledge Works model.[2]

Yanagicho's intrapreneurial success is well documented. Its most recent forays into optical-electronic devices, IC cards, and nonimpact printers represents a leap forward in capital intensivity and technical capabilities. Yet it is a leap that is clearly rooted in past accomplishments. Traditionally, product and process advances at Yanagicho have been evolutionary, building on past achievements. However, products today are much more demanding of research, development, design, and manufacturing skills, and so Yanagicho has become a fulcrum for leveraging resources gathered internally as well as from other factories, divisions, affiliates, and suppliers. Rather than simply adapting, Yanagicho is fusing technologies, products, and processes, innovating in almost revolutionary ways.

The challenge is developing families of attractive, related products on time, with well-orchestrated marketing campaigns. Time to market is critical in this challenge, and increasingly time to market is based on the amount of time spent on incubating new technologies, product design and development times, time spent managing suppliers, production lead times, and production setup times. Time savings in all these areas are the very essence of Knowledge Works. Toshiba's managers are constantly echoing the theme, "market in, not product out," which is to say that a lot of time, money, and talent can be wasted on products for which there are no markets. Markets dictate what, how, and when to make, an admonition taken to heart in Knowledge Works (although not without an occasional hard reminder; witness the SuperSmart card project).

People and Intellectual Capital

There are also human resource reasons for why manufacturing sites like Yanagicho succeed. Most simply, it is an exciting place to work. The breadth and depth of the work attempted; the challenge of competing against other departments in bringing new products on line; the drive toward education and constant improvement in performance; the deepening of experience, awareness, and cooperation; the excitement of working with highly trained and involved colleagues—for these and other reasons, Yanagicho attracts and motivates employees.

The importance of education is manifest throughout the factory. It is evident in morning roll-call when the day's activities, including educational activities, are detailed in hundreds of work teams and sections. It is evident in QC circles as well as in dozens of on-the-job training programs. Off-the-job education receives no less attention, and fully one-third of the budget of the General Affairs Department (the largest and most important staff function at Yanagicho) is devoted to education. In the early 1990s, Yanagicho was offering 8 different categories of off-the-job training; given that there were eight to ten different classes in most categories, the range of educational offerings is really quite remarkable. In-factory education plus a constant need for more on-the-job training push the factory's capacity to upgrade and enhance its human resources. The attention given to human resources, once good people have already been hired, underscores the importance of Knowledge Works as sites of intellectual discovery, creation, and application.

Without manufacturing facilities like Yanagicho, Toshiba would not be able to compete with rivals like Matsushita Electric Industrial, Hitachi, and Mitsubishi Electric. Toshiba is already hard-pressed to compete across the broad product range that it does. The capabilities to compete against worthy rivals across a product spectrum with speed, flexibility, and innovation are inherent in the organizational design and people of Knowledge Works. But such capabilities are fragile.

In order to move easily between functions, products, and processes, the depth of capabilities in any one function, product, or activity should not be too great. The aim is to nurture flexibility in a range of activities that is at once broad enough to absorb the risk of market failure in one product segment yet narrow enough to maximize economies of learning by sharing resources and capabilities across somewhat similar products and operations. Focused flexibility—a somewhat ambiguous concept—is the goal.

An alternative to the term *focused flexibility* is the concept of "integrity." As used in the management literature on product development, integrity means the ability to develop products that take maximum advantage of available resources and capabilities, and result in products that make sense, correspond closely with what customers want and with what companies can offer.[3] The delivery of products that make sense in the light of innumerable and constant trade-offs suggests an ability to take effective action or to act with integrity.

While in some respects designing photocopiers or laptop computers may resemble product development activities in the motor vehicle industry, it is much more difficult to satisfy and, perhaps, to predict the behavior of institutional users as compared with individual consumers. Today's office automation market is filled with a bewildering mix of interactive electronic gear and competing technologies, and simply trying to stay in the market is daunting enough on its own. In this sense, Yanagicho's products do not pass the same sort of integrity test that, say, Honda's products do. Yet Yanagicho's system of focused flexibility makes products that pass many reliability, quality, availability, and utility tests, which altogether constitute a kind of integrity check.

Flexibility enables very short production runs, such as the hundreds of different electric motors among the 11,500 different kinds of products made by Toshiba in 1992.[4] By encouraging flexible shop floor practices and empowering bottom-up organizational processes, factory workers respond to the market as well as to management. They obviously work within rules and routines laid down by managers but these wisely urge shop floor participation and teamwork as critical pinions linking factory-based capabilities with market demand. A lively, vibrant, market-oriented shop floor culture, in this sense, is the key to manufacturing ambition.

The strength to pioneer markets and successfully compete against rival firms carries great risks. A tightly clustered product strategy, where a limited number of highly sophisticated products are made for specialized markets, can be abruptly derailed by innovation breakthroughs elsewhere. Risks can be minimized, however, by broadening and deepening technical and organizational capabilities at the factory level of organization and by creating a knowledge domain where both focus and flexibility may be realized. Yet, focus, flexibility, and versatility embodied in committed human resources, workplace-centered manufacturing strategies, networked resources, and product/process innovation are not acquired easily, quickly, or fortuitously.

Nor inexpensively. A willingness to hold together the functional requirements for strategic flexibility in the face of high capital investment requirements as well as high land and labor costs is remarkable, and that willingness distinguishes Toshiba from many Japanese and most Western rivals. By financial measures alone—capital return projections, investment hurdle rates, and depreciation allowances—the Knowledge Works model is a risky one. But by some measures of profitability, such as the ratio of cash generated during any given year to investments made in new and existing technologies in the previous year, Knowledge Works can reap remarkable rewards.[5] This happens when one or several product lines are big winners, especially under Tokyo Effects and in global markets. Then, the advantages of investing in people by creating, transferring, and sharing intellectual capital in multiproduct, multifunction plants are manifest.

Knowledge Works and Path Dependencies

The antecedents of Knowledge Works or what have been called focal, lead, "mother," or "parent" factories are found in the prewar period.[6] Focal factories served as sites of territorial administration when higher-level, companywide ordering of production and distribution proved difficult. They organized production, planned marketing and distribution, and generally coordinated a host of value chain activities on local and regional levels.[7] In doing so, they were also the primary vehicles for transferring Western technology into local practice and products. After the war focal factories were joined with new personnel and management policies and with flexible production routines, small group activities, elaborated hierarchies, and other features of modern manufacturing organization.

So, Knowledge Works appeared as institutional responses to a set of historical contingencies, such as bureaucratic failure in the prewar era, extremely high rates of postwar economic growth, especially from the 1950s to 1970s, rapidly shifting markets, dramatic improvement in engineering and manufacturing know-how, and the intensification of rivalries across a broad spectrum of product lines. A managerial riposte to these contingencies was to consciously choose to endow certain systems of manufacturing with special qualities: sufficient resources to respond systematically to new market and technological opportunities and an urgent mission to do so, quickly and repeatedly. As these choices and actions were repeated and accumulated, human experience piled atop human experience, site- and relation-specific institutional values and practices became possible.[8]

Corporate Strategy and Factory-Based Innovation

During the past 30 years, Toshiba has stood its product lineup on its head. The traditional order of importance was heavy electrical equipment, light electrical and household goods, communication and information devices. Today, the order is completely reversed. Toshiba did this by emphasizing applied research, product/process innovation, supplier relations, and time-to-market speed—the Knowledge Works model, in effect.

From 1955, four waves of new products have powered Toshiba's sales growth. The first three of these came from traditional product lines. From 1956 to 1965, electric power generation and transmission equipment and black and white televisions were the core products pushing Toshiba's average 33 percent growth in revenues. From 1965 to 1971, color televisions and atomic energy plants moved revenues along at a 46 percent rate of annual growth. Then, from 1971 to 1986, a 15-year hiatus, Toshiba struggled along at a lackluster 10 percent rate of annual revenue growth based on the export of an aging portfolio of household appliances and electric power equipment.[9]

The growth of the 1950s, '60s, and '70s required firms to reorganize themselves in anticipation of demand. At the factory level of organization, product variety increased and production volume rose significantly. Yet focal or lead factories retained their administrative importance, constraining but not impeding a gradual process of corporate divisionalization. Since the late 1970s, manufacturing strategy has been less concerned with volume and more with product variety and time to market. To accelerate the flow of technology to the factory floor and, hence, to boost product/process innovation and improve time to market, in 1977–78 Toshiba conceived an entirely new sort of R & D organization. Rather than centralize R & D in corporate labs at the top of the hierarchy, engineering labs were collocated in factories.

In 1978, a Nuclear Energy Engineering Laboratory and a Consumer Electronics Engineering Laboratory were founded. In the next year, a Semiconductor Engineering Lab and in 1980 one for Electronic Devices and another for Medical Equipment were opened. An Information Systems Engineering Laboratory and a New Materials Research Lab were established in 1984 and 1987, respectively. By

1987, Toshiba boasted seven engineering labs, all collocated with factories, and another seven corporate labs. Subsequently, two of the corporate labs merged, and half of the remaining six were resited at manufacturing facilities. In other words, nearly three-fourths of Toshiba's R & D assets were sited low in the corporation, collocated with manufacturing facilities.

The results were entirely predictable and desirable. Speed of product development increased, costs of product development fell, and, most important, information and knowledge flows were enhanced up and down the value chain. Most notably, Toshiba was way ahead of the competition in this organizational innovation, especially after 1987 when two more engineering labs were sited with factories. But the effects and advantages of Knowledge Works are apparent to rival firms. Mitsubishi Electric moved toward the Knowledge Works model by concentrating more of its applied research and engineering resources at Itami Works near Osaka in 1993, and Hitachi initiated several factory-based, large-scale R & D projects in 1994. In comparison with Toshiba, however, they were some 15 years late to the party.[10]

The collocation of engineering labs in factories and the appointment of Shoichiro Saba as CEO ended the postwar era for Toshiba. In 1987, Saba launched his "I" strategy. "I" comes from the words *intelligence, information*, and *integration*. Saba's strategy was an all-out shift of corporate resources toward communications and information technologies (IT) where Saba figured that Toshiba could excel. Saba's "I" strategy represented a major break with Toshiba's past products, and it came none too soon. The rising value of the yen, especially since the Plaza Accord of December 1985, put severe price pressures on the competitiveness of Japanese products.

From 1985, two-thirds of Toshiba's R & D and capital improvement expenditures have been targeted in communication and information technologies. Toshiba's organization was revamped to better accommodate such huge shifts in internal resource allocations and to respond more quickly to the diverse needs of communication and information technology customers. This was accomplished, as we saw, by siting engineering labs in factories, and by developing robust networks of suppliers, by keeping design, development, and tool-making skills together with product/process engineering, and by improving organizational change techniques, like TP campaigning.

Lest one think that this shift in emphasis was a foregone conclusion, remember that Toshiba was no more oriented toward communications and information technologies than either Hitachi or Mitsubishi Electric, the other general electrical equipment/electronics makers. Moreover, Toshiba's success with semiconductors, computers, and industrial and office equipment products came hard on the heels of a number of earlier failures, in part because Toshiba inherited from General Electric a slew of products and technologies that were far from cutting-edge. Given a traditional reliance on heavy equipment and appliance sales, less than state-of-the-art technology, and limited experience with integrating basic and applied research in engineering and development processes, Toshiba understandably stumbled a number of times along the way.

Toshiba turned the corner on its "I" strategy when it was first to market with an electronic typewriter, called Rupo. Chinese characters are used in writing Japanese, and traditional mechanical typewriters set thousands of characters like offset printers handset type, obviously a cumbersome and costly process. Rupo simplified, improved, and accelerated communications by allowing Chinese characters to be written electronically. Sales were astonishing. Within a few years, hundreds of thousands of Rupos and Rupo look-alikes were selling. Once letters could be typed electronically, it was not long until they could be stored, manipulated, and retrieved in the same way. Electronic typewriters became word processors and personal computers.

The cost of products like these vary directly with the cost of semiconductor memory devices. A useful rule of thumb during the 1980s was that the cost of semiconductors fell by half while output doubled every year. Fourfold annual improvements in cost-performance ratios brought electronic goods within everybody's budget. The market for smaller, lighter, quicker IT devices seemed limitless, and Toshiba was in the thick of it.

But then disaster hit. While Toshiba claimed no direct knowledge of Toshiba Machine's sale of sensitive machine tool technology to the Soviet Union, both its chairman and president resigned. The resignations were intended as a way to accept formal responsibility for the machine tool sales and to lessen rising American complaints about how much damage had been done to U.S. strategic capability. But the resignations were misinterpreted as admissions of guilt and responsibility. On Toshiba's side, they were only admitting organizational responsibility for one of their subsidiaries. Predictably, the resignations heightened rather than lessened criticisms.

Toshiba products were banned from American government purchases for three years. A huge order of some 80,000–90,000 laptop computers for military use was lost as a result. Internally, Toshiba spent millions installing a compliance program that trained employees and set up a monitoring system for auditing strategically sensitive sales (affecting smart card development as seen in chapter 6). Externally, Toshiba launched a lobbying effort that still draws the ire of Washington insiders who dislike foreign firms that best American firms at their own game.

Toshiba also redoubled its efforts to pull away from the pack by designing and developing higher-value-added products. The surest way of doing so was to offer new and better products faster than rival firms. Toshiba did so in spades. Newer, faster laptop computers appeared almost monthly. One and 4 megabyte DRAM chips enjoyed a huge demand. Dynabooks—booksize portable computers, smaller and lighter than existing laptops—took the market by storm after their introduction in July 1989, catapulting Toshiba's earnings and sales ahead.

Out with "I" and In with "W"

In 1988, former president and chairman Saba's "I" strategy was supplanted by the new CEO's "W" strategy. Junichi Aoi came up the traditional way, as a heavy electrical equipment engineer. Yet, on his elevation to the presidency, he leap-frogged Saba's concentration on intelligence, information, and integration with a new vision. "W" stood for the world, and Aoi's goal was to globalize Toshiba in three ways:

1. More strategic alliances and joint ventures with foreign firms, indicating that the Toshiba Machine debacle had taught the value of having foreign allies.
2. More transfer and implantation of production capacity overseas, particularly in Southeast Asia and China.
3. More imports and sourcing from overseas, mostly of electronic materials, parts, and components.

Aoi's "W" strategy was foreshadowed by an unusually positive and productive relationship with IBM. The story begins in the 1960s when Toshiba abandoned its own march to develop and produce mainframe computers. Thereafter, Toshiba divided its computer purchases among a number of major vendors, including IBM. By the mid-1980s, 25 percent of Toshiba's distributed computing capacity was based on IBM equipment. Given this and that Toshiba's main rival in IT technologies was NEC, a non-IBM-compatible company, Toshiba decided to standardize on IBM operating systems.

So Toshiba joined Hitachi and Fujitsu, two firms already on the IBM standard, as it locked horns with NEC, a company that was enjoying a 70 percent domestic market share in personal computers. Toshiba quickly converted tens of thousands of Japanese citizens to the IBM standard with the Dynabook's open architecture, its power, portability, third-party software, and IBM compatibility. By the early

1990s, NEC's market share had slipped to under 50 percent, handsomely benefiting Toshiba and IBM.

One good turn deserves another. In November 1988, Toshiba and IBM (Japan) formed a joint venture, Display Technology, Inc., for manufacturing liquid crystal displays. About the same time, Toshiba inked semiconductor technology licensing agreements with Motorola of the United States, Philips of Holland, and Siemens of Germany. With Sun Microsystems, Toshiba agreed to use Sun's RISC (reduced instruction set chip) technology in a portable engineering workstation. With NTT (formerly Nippon Telephone & Telegraph), Toshiba developed bank card systems, ATMs, and PBXs (private branch exchanges), and by doing so, Toshiba wedged itself between NEC and Fujitsu, both traditional NTT suppliers. NTT could not ignore Toshiba's rising star, and they have continued a mutually profitable relationship during the 1990s.[11]

Capitalizing on "W" Strategy

From late 1989 until well into 1992, Toshiba laptop and dynabook computers were selling in the neighborhood of 75,000 units per month. While facing pricing pressures from models made by rival Japanese, Korean, and Taiwanese firms, Toshiba has been able to maintain a lion's share of the high-end market with a reputation for quality, IBM compatibility, and good value. Toshiba also enjoys an enviable record of convincing its laptop customers to trade up to higher-end models. With a dozen different models to choose from, consumers enjoy a cornucopia of price-feature-power trade-offs.

With about 75,000 computer units being assembled monthly, increasing returns to scale as well as significant learning economies can be expected. Not only are there scale effects in materials sourcing and logistics but, more important, learning curve effects can drop costs per unit by perhaps 5–10 percent per quarter. In tandem, scale and learning economies are saving Toshiba billions annually. Lower market prices and higher labor costs may offset some of these savings, but lower unit costs and higher profits are still the end result.

Laptop computers are not alone. Semiconductor memory prices have been falling an average of 13 percent a year; other semiconductors at 5 percent a year. Driving costs down on these and other products depends on three interrelated organizational skills and capabilities. All of these are what Knowledge Works are designed to capture and exploit. The first is straightforward scale economies but, as always, scale economies infused with learning effects are important. Given many laptop models, production runs for any single model may not be so large. For example, an active color matrix, 586-

chip laptop may be produced in as few as several thousand units per month. But the process technology and many higher-end parts and components are shared with lower-end models, so scale-oriented learning across a broad band of similar products is predictable. This allows Toshiba to add capacity ahead of demand because experience has shown that volume production will necessarily drive down costs.

Second, value analysis (VA) techniques can be an important source of cost savings, especially when it is realized that as much as two-thirds to three-fourths of the value of high-end, IT products may be sourced from suppliers. Reducing lead times while maintaining quality is clearly the key issue with suppliers. Well before manufacturing, product and process standards have to be fixed, supplier resources mobilized, and the latest technologies incorporated in design and development. VA techniques are applied by assemblers as well as suppliers, and complementary learning and cost savings are expected in such instances. Even though VA savings are not always distributed equally, mutual problem-solving and profit-taking opportunities seem to abound. (Recall chapter 4's story of assembler-supplier relations.)

Third, Toshiba's factory-based system of applied research, product development, and product/process innovation—the Knowledge Works model—takes full advantage of learning opportunities inherent in the manufacture of IT products. As more and more power is crammed onto chips, devices are increasingly miniaturized, leading to product design and development processes that require high-level systems integration capabilities.[12] Not surprisingly, such product/process design and systems integration capabilities are site- and relation-specific.

The histories of Ome, Yanagicho, and other Knowledge Works affirm that institutions can think, learn, evolve, and create distinct, site- and relation-specific competencies. Seminal work in the areas of institutional culture and theory seems to find ample confirmation in Yanagicho's story.[13] For more than a half-century a kind of ameliorative and progressive force has flourished within Yanagicho. The stimuli for advancement have been many, and uncountable in a real sense. Knowledge Works were the means for implementing Aoi's "W" strategy, and today they embody Toshiba's competitive edge in making high-value-added products.

The lessons of capitalizing on organizational knowledge have been learned elsewhere but nowhere more effectively than in Japan and by Toshiba. In the case of General Electric, for example, a great many production facilities, including research and development activities, have been localized in Schenectady, New York. But as GE prospered and as new product divisions proliferated after World War II, the facilities and functions that remained in Schenectady tended to be older product lines and more traditional manufacturing activities. Plants incorporating newer manufacturing techniques and the accompanying higher-tech products were sited outside the traditional manufacturing zones of New England. Separating old sites from newer technologies, products, and processes and, at the same time, separating newer plants from the accumulated knowledge and know-how of older sites lessened GE's industrial competitiveness.

In the cases of Dupont and 3M, two other American industrial giants, laboratory-factory facilities are located near the company headquarters in Wilmington, Delaware, and Minneapolis, Minnesota. However, the facilities are designed to bring products along to the point of trial manufacture only, and in this sense they remain facilities for experimental production. Prototype factories and one-off factories such as these do lower the costs and speed up processes of new product development but often they lead to organizational bottlenecks farther down the manufacturing and marketing value-chain.

By combining the old with the new, however, Yanagicho represents an organizational strategy for coping, adapting, and staying ahead of the competition. The factory knows what to know and how to know or, said less grandly, how to adapt effectively to internal and external change. Yet coping and adapting do not occur in a vacuum. Motivations are closely linked to social and material rewards. Employee suggestions are rewarded monetarily. Workplace remuneration is tied to teamwork, productivity, and attitude. Bonuses and promotions are related to performance. Yanagicho did not begin with this strategy; it evolved, somewhat naturally and somewhat consciously. As one Yanagicho engineer told me, "What we leave behind for tomorrow is more important than what we're doing today."

Knowledge Works and Tokyo Effects

Rates of social and economic change appear to be accelerating everywhere, but perhaps nowhere more so than in Japan. Because of Tokyo Effects, rapid rates of change and Knowledge Works are interrelated. Thermodynamic laws suggest that when molecules are compressed, they become excited, causing electrons to race faster and gain temperature. Tokyo Effects are analogous to thermodynamic laws in that increasing rates of change in Tokyo (as before, I am referring to a metaphorical Tokyo encompassing the megacity corridors of central Japan) force organizations to respond faster, communicate and coordinate better.

But the costs of locating manufacturing facilities in megacity areas and of staffing, managing, and training skilled employees there have soared. Land in Tokyo is among the most expensive in the world, even after the recent drop in values; land costs in Osaka, Kyoto, Kobe, and Nagoya are not far behind Tokyo. Across-the-board increases in costs, rates, and the pace of change are not viewed dispassionately by Japan's firms.

With few natural resources, limited surplus labor, and relatively expensive land, Japanese firms are forced to increase their value-added output by developing more efficient and varied uses for existing resources. Most firms did not have sufficient scale of activities to invest heavily in central R & D labs until recently, so investments in sales engineering centers, engineering labs, and other pre- and postproduction units, were concentrated in the existing factory-based system of technology management. More centralized investments came, of course, mostly during the 1970s and 1980s as firms built up their divisional and corporate facilities. But central R & D labs came late to Japan (relative to the same phenomenon in the West). They came after focal factories had already developed robustness and range in manufacturing and managerial activities; they came with less breadth and depth in research activities than in comparable Western firms. For such reasons, corporate Japan's response to a postwar speed-up of scientific and technical change and to an emergence of global high-technology industries was distinctive: *greater investment in an existing enterprise system effectively linking factories, firms, and interfirm networks in R & D, design, development, production, and distribution activities.*[14]

Thus, Knowledge Works were an adaptive, evolutionary response to Tokyo Effects and the new time-to-market competition. However "new" in their most evolved form, Knowledge Works are grounded in Japan's industrial history, that is, in a tradition that emphasizes organizational learning, full utilization of and flexible investment in human and material resources, and market responsiveness. They evolved in response to a series of trade-offs: organizational differentiation and integration, specialization and flexibility, routine and nonroutine operations, competition and cooperation (with affiliates and suppliers). The points of balance were less important that the notion of balance, since rapidly changing circumstances make implausible anything but temporary moments of balance.[15]

At Yanagicho, the strategic response was neither product nor market in any narrow sense. Instead, dynamic balance was pursued by advancing two sorts of technologies—traditional ones of metal cutting, precision machining, toolmaking and die-casting, and newer, government-encouraged technologies of pattern recognition, optical- and microelectronics. These were blended with a "can do" attitude based on a 60 year history of successful manufacturing, mostly cooperative labor-management relations, macro-organizational planning, budgeting, and coordinating skills at the top of the factory. Design, development and product/process competencies accumulated at the product departmental level.

But firms are always constrained by an available mix of people, capital, equipment, and know-how. Knowledge Works cannot perform at peak levels for long, and trade-offs must always be made. No matter what capabilities Knowledge Works embody, managers shape, mold, and move those capabilities in anticipation of change. When managers make the right choices, the combination of Knowledge Works-like strategies and thermodynamic-like compression of time, space, and purpose *may lead* to capabilities that are nothing less than spectacular:

- capabilities to commercialize two or three times as many high-value-added products as rival firms,
- capabilities to incorporate two or three times as many technologies in new products,
- capabilities to bring higher-value-added, technologically complex products to market in less time,
- capabilities to compete across the board: not only in established product segments but also to pioneer the commercialization of new products and markets.

The juggling/balancing/mixing/channeling of these capabilities under Tokyo Effects first appeared during an era of especially rapid economic growth, 1959–1973. Beyond this period of intensive growth, runaway consumer spending at home with reopened markets overseas pushed the Japanese economy ahead at an astonishing pace for an entire quarter-century, from the time of the Korean War to the mid-1970's oil crises. Across-the-board widening and deepening of markets coupled with a monumental transfer of "best-practice" technology from abroad pushed firms to explore new notions of manufacturing organization, which, moment by moment, profoundly transformed manufacturing strategy.

Knowledge Works were a result. And they were a cause, once firms realized the competitive advantages they conferred. These advantages culminate in positive workplace attitudes and creative employee relations. Employees are pulled together and given a sense of specialness, commitment, and teamwork. High levels of self-motivation and work group autonomy result. Without these, the tasks of integrating a dozen different product lines, scores of interdependent functions, several hundred parallel streams of parts, components, and subassemblies, not to mention thousands of different employees and suppliers, would be overwhelming.

This is a social innovation, sufficiently dramatic to suggest something akin to a "shop floor revolution," where line employees are the foci of innovation and ingenuity. To be fair, however, "shop floor" should be broadly inclusive of all employees who are directly and indirectly involved with factory-based R & D, design, development, and production.[16] The most extraordinary feature of this "ordinary" factory is Yanagicho's seemingly impossible integration of all this ambition, variety, and complexity with speed and reliability, culminating in four people-dependent competitive advantages.

Product/Process Innovation

Product innovation at Yanagicho comes from two principal sources: the close alignment of all industrial R & D and manufacturing functions in one organization; and in the proximity of numerous product lines and associated assembly technologies, so that concepts, techniques, and experiences from one product stream can be shared with others. The first of these sources of innovation creates numerous positive feedback loops in design, development and manufacturing. The latter maximizes the advantages of economies of scope and learning due to a sharing of personnel, physical distribution facilities, technical and production know-how.

As a case in point, the newer products manufactured at Yanagicho can be traced to two innovative projects initiated in the mid-1960s: the development of a prototype mail sorting machine for the Japanese Post and Communications Ministry in 1965 and, in 1966, the production of electronic typewriters (Rupo), which were originally developed by the Toshiba Type Company. Both devices employed sophisticated electronics, optical character recognition (OCR), paper handling, sorting and printing technologies that now characterize a whole range of Yanagicho's labor-savings devices, from cash handling and ATM machines to the automatic fare collection system for the Massachusetts Turnpike Authority, as well as photocopier and laser beam printers.

Speed and Cost of Product Development

The speed and costs of product development, like product innovation rate, basically spring from the same proximity, close alignment and integration of all the resources needed to conceive, design, develop, refine, make, and market products in one place. In markets where speed and flexibility are everything, reducing the time and costs of product development while building in quality, reliability, and performance are absolutely essential.

In Japan today, the product life-cycle for PPCs is a half-year; that is, within six months a rival firm will bring out a competing PPC model with greater functionality and often lower cost. Although particular PPC models may stay on the market for 18 to 24 months, they are passé in half that time. In order to stay competitive, technicians and engineering specialists from many functional areas have to work closely and intensively together.

The ability to walk a short 20 meters from Yanagicho's Works Laboratory, where research on products and processes with a three to five year time horizon is conducted, to visit product design, tool and die, and manufacturing engineering departments, or to view prototype mock-ups on the shop floor, imparts immense design and development advantages. Employees in one functional area know and work cooperatively with those in others. As a result, the time and cost of developing new products and of enhancing existing ones are considerably reduced. Also, when necessary, collocation of R & D and production activities allows people to move with products. Bright ideas and those with them can move to engineering labs, and once abstract ideas are brought to the point of production feasibility, they can move again to product departments and the shop floor. Good ideas, good researchers, and good manufacturing practices are often found together.

Productivity

Productivity is another consequence of Knowledge Works. By whatever measure, such as value-added output per employee or the rate of capacity utilization, Knowledge Works perform well. Of course, productivity may be higher in mass production factories when the output of a single line at a single point in time may be running at a 100 percent capacity or better. In Knowledge Works, production runs and, thus, capacity utilization rates vary considerably. But Knowledge Works' wider range of products and models may allow for more stable rates of capacity utilization, especially in turbulent markets, and for greater reuse and recycling of existing resources. Output per unit of investment is likely to improve over time.

For instance, if demand in one product area drops markedly, it is common to redeploy resources, such as line workers or parts suppliers, to another area with growing, or at least stable, demand. Additionally, the start-up costs associated with getting a new model or product line into production can be offset by the established revenue streams of existing models and products. Yanagicho's nonimpact printers (NIPs), for example, are part of the business group producing plain paper copiers, PPCs. Since building market share is a time consuming process, it is less expensive and more convenient, from a cost accounting point of view, to situate NIP with the PPC group. When costs of product development and manufacture are considered in this way, productivity, reusability, and speed of development are all likely to be higher in Knowledge Works than in mass production factories. Product line and factory productivity gain as a result.

Employee Involvement

Another advantage of Knowledge Works may be found in their higher levels of employee involvement. The reasons for higher involvement are many. Larger numbers of product lines permit greater job rotation, enlargement, and enrichment. Since product and process innovations are explicit goals, great importance is placed on creativity, collegiality, and considerations of how to use less energy, time, and materials while on the job. Opportunities to do different jobs, rethink work content, and train for new positions motivate. Such opportunities almost always require greater work involvement, which results, in most cases, in higher levels of participation and commitment.

Of course, higher levels of involvement may backfire if ample opportunities for fulfillment and satisfaction on the job are not available. That too drives the cycles of organizational change and renewal found in TP campaigning. And quality assurance managers, for example, are asked not only to build quality into the various products being made and assembled in the factory but, just as importantly, to manage the culture of work satisfaction and reward on the shop floor. Committed and involved people make good products, and not vice versa.

In this way, mental and psychological aspects of employment are tied tightly to material rewards, social incentives, and a positive cycle of product and process innovation. It is recognized that customary effort norms, based on what are considered acceptable levels of work performance, are a source of what has been called the "productivity/technology dilemma."[17] In the opinion of many at Yanagicho, high employee involvement springs from a conception of work as culture, and a recognition of the need to try and manage organizational culture. A big part of managing culture, as seen in TP campaigning, is directed against crippling customary effort norms and substituting more challenging and rewarding norms in their place.

The tug of so many different trade-offs and a countervailing need to train versatile, committed, and willing employees can be graphi-

	Organizational Units		
	Divisional Affiliated Units	Departments	Sections
Factory Units			
General	7	7	18
Information Systems	10	10	35
Household Appliances	1	1	3
Instrumentation & Automation	2	2	9
Resident Units			
Software Research Lab	3	0	13
Central R&D	3	0	3
Air Conditioners Compressors	0	0	1
Materials Research	1	0	2
Information Systems Technical Center			
Patents	0	0	1
Manufacturing Engineering	0	0	3
Service	0	0	4
Office Automation	2	0	6
Business Machines	2	0	13
Peripherals	1	0	1
Research	5	0	25
Totals	**37**	**20**	**137**

Figure 7.1   Structuring the workplace at Yanagicho.

cally demonstrated by underlining the organizational complexity of Yanagicho. As seen in Figure 7.1, 37 different divisional level units have organizational ties to 20 departments and 137 sections, spreading the efforts of Yanagicho's nearly 3,300 regular and resident employees over several dozen different operational, managerial, and technical departments. Another 10 units affiliated at the divisional level but resident at Yanagicho are found at the Information Systems Technical Center housed in the Power Tower. Adding part-time and temporary employees, including a permanent if rotating complement of "visiting engineers" from suppliers and affiliates, further enriches the picture.

Having detailed some consequences of Knowledge Works organization and strategy, it is useful to emphasize again how unorthodox the model is, at least compared with conventional operations management thinking.[18] Hence, Knowledge Works represent a new model of manufacturing organization and strategy, delivering first-to-market performance, creating knowledge and know-how, and accumulating skills and competencies across a full range of applied research, product design, development, and manufacturing requirements. Yet Knowledge Works are not skunk works, prototype factories, or job shops for one-off production. They do not produce for NASA, Europe's ESPIRIT, or some MITI-sponsored, future-oriented project. They are factories, not hush-hush laboratories or one-off design centers, developing and making reasonably priced products for discerning yet ordinary consumers.

Creating, Innovating, and Adapting

The technical and organizational strengths of Yanagicho take form in more than a dozen different product lines, from custom order to mass production goods, encompassing almost every category of manufacturing engineering technology. Applied product and process research, design engineering, prototype development, model and full-scale manufacturing, as well as some marketing and sales are all conducted and managed on site. It is almost unheard of that so many technologies, processes, and products are combined in single manufacturing sites.

The orthodox assumption is that the routinization and specialization requirements of high-volume/low-complexity production necessarily conflict with the flexibility and functional integration requirements of low-volume/high-complexity manufacturing. In other words, it is difficult to realize focus, quality, reliability, and above all, speed in widely varying manufacturing circumstances. Yet the record of Yanagicho suggests otherwise. Skills and resources can be shared across widely varying manufacturing circumstances as long as there are well-defined pathways as to what and how knowledge will be shared. Those pathways are found by the physical, technical, social, and organizational boundaries that together describe the parameters of what happens and how it happens at Yanagicho.

Yanagicho involves even more than new products, advanced technologies, and complex organizational arrangements. Knowledge Works are a significant social innovation, one where superior human resources are well organized, well managed, highly motivated and productive. Knowledge Works are the culmination of a factory-centered approach to technology transfer and industrial development, and of Japanese attitudes toward workplace culture as learning (*noryoku*), effort (*doryoku*), and cooperative enterprise (*kyoson-kyoei*). But Knowledge Works are more than a simple accumulation or culmination of these qualities. These attributes are embodied in Knowledge Works through a conscious effort to work, learn, and share together, and however imperfectly, to manage that process. Managing intellectual capital more effectively is the Knowledge Works strategy.

These national, corporate, and cultural features of an industrializing Japan coalesced during an era of remarkable economic expansion worldwide, and found their full expression in factories like Yanagicho. Knowledge Works are clearly instrumental in pushing Japan into the forefront of industrial discovery, rapid product development, and effective process innovation. They will become even more crucial as Japan is challenged by protective legislation on one hand, and rising competition from neighboring Asian economies on the other.

The history of Yanagicho, therefore, is a history of technical innovation and organizational adaptation. Plus, it is a story of human ambition and effort exercised in the interest of accumulating manufacturing skill, insight, and experience. In its details, it is a highly particular history of one plant and the people who worked there. More broadly, the lessons are general. The Yanagicho Works offers one system, one solution if you like, to the problem of making many different attractive objects in reasonably large volumes. Yanagicho's strategy of making only what you need, no more, is modeled to a degree on Toyota's production system of just-in-time manufacture.

Managers at Yanagicho, however, recognize that Toyota's system will not work equally well in every industry and plant. Although Toshiba's products are in general far simpler than Toyota's, the number of common parts per model or unit of production is far higher for Toyota than for Toshiba. This is because Toyota's production runs are much longer than Toshiba's and because Toshiba offers a much greater variety of product models, shapes, and sizes than does Toyota. Also, in Toshiba's case the frequency of design changes is higher—every six months for plain paper copiers, for example—and a higher frequency of design changes impacts the entire research, design, development, and production process. When product life-cycles are as short as six months, it is difficult to produce in small lots with many variations *and* without significant in-process inventory.

There are significant differences in terms of how Toyota and Toshiba are internationalizing their production systems. It is possible for Toyota to design and build an new entirely new motor vehicle from scratch in the United States in spite of all the complexities of automobile design and manufacture. Yet, designing and building an entirely new photocopier in the United States is something that Toshiba is not yet able to do and, perhaps more significantly, is not likely to attempt anytime soon given a need to concentrate PPC research, design, and development resources in Japan where competition is fiercest. The internationalization of Toyota's capabilities, as seen in Toyota's successes at NUMMI in California and Georgetown, Kentucky, suggests different production and process requirements for motor vehicles as compared with electronics/electrical equipment, as well as greater willingness, need, and experience on Toyota's part to produce overseas.[19]

Given differences in motor vehicle and industrial electronic product markets, it is difficult for Toshiba to apply Toyota's production model, even though many Toshiba managers may want to do so.[20] Moreover, the Toyota system is rooted in organizational learning and, as such, it is a model of how things may work, perhaps should work, after considerable on-site adaptation, experimentation, and adjustment. Organizational learning is essentially a process of on-site cultural differentiation, so what works well for Toyota will necessarily not work as well for Toshiba.

Toshiba's deployment of human resources in favor of the Knowledge Works model are revealed in Figure 7.2. Of Toshiba's main 27 main factories (3 more are branches factories), about half have particularly high ratios of indirect to direct workers or high ratios of knowledge workers (managers, designers, engineers, and researchers) to manual workers. Among these, Yanagicho has a high ratio of managers to other sorts of indirect workers, attesting to the diversity of Yanagicho's product lines and product support functions. (Yanagicho is the number ten bar in Figure 7.2.) In fact, only 36.4 percent of all regular employees, regardless of education level, are production workers at Toshiba, and adding regular and resident employees, 90 percent of Toshiba's 19,000 university trained scientists and engineers are sited in Knowledge Works.

Knowledge Works are available for everyone to see and appreciate and, quite obviously, the complexity of their form implies that they were designed for a purpose. Mitsubishi Electric, for example, distinguishes between "brain" factories and "work" factories.[21] But in order to exploit a strategy of factory-based R & D, a foundation of adaptive, reflective experience in product/process innovation and systems integration must be built. While Knowledge Works offer a model of how to do this, newly industrializing

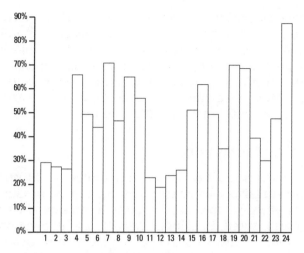

Figure 7.2 Knowledge Workers at Toshiba factories, as a percentage of all non-production, non-administrative regular employees at 24 main Toshiba factories in 1988–1989.

and reindustrializing countries cannot readily imitate Toshiba's Yanagicho Works. Yanagicho's capabilities with respect to PPC design, development, and manufacture are singular and not at all imitable, one to one.

In important ways, therefore, what is happening at Yanagicho is not what is happening elsewhere. Elsewhere, companies are responding to perceived crises, challenging economic conditions, disappearing markets, and redundant workers, scurrying about trying to find solutions. What Yanagicho is doing is what Yanagicho has always done. With nerve, skill, and a knack developed over 60 years of evolution during difficult prewar, wartime, and postwar conditions, Yanagicho is making adjustments.

The history of how the Yanagicho Works got from there to here is a history of effort, risk-taking, and determination. More concretely, it is a story of how incremental but continuous efforts of adaptation, experimentation, and dedication have paid off in employee involvement, learning, and the growth of organizational knowledge. These resulted in a generation and integration of synthetic energies and capabilities; quicker response times to market opportunities; an increase and intensification in research and development activities; a reduction of formal hierarchy as a means of organizational control; a renewal of human and material resources.

In sum, Yanagicho is used to creating and adapting. Things are very different now compared to 5 years ago, and greatly different compared to 10 years ago. Innovating in the midst of adapting is really Yanagicho's most important skill and capability—its core competence and essential intellectual capital. This is a cultural asset in that learned patterns of adjusting to markets and technical change and of adapting to other business units inside and outside the firm, have become institutionalized in a rich, thick factory culture. Organizational culture gives Yanagicho the ability to cope, adapt, and improvise in the midst of uncertainty and risk.

Knowledge Works in Review

Knowledge Works is about manufacturing strategy and Toshiba's way of managing its factory-based resources and capabilities since World War II. Toshiba's way favors full-function, multiproduct factories that emphasize speed, learning, creativity, and innovation up and down the value chain. Toshiba's central location in greater Tokyo, an epicenter of industrial marketing, design, research, and development, has had an important effect on the evolution of Knowledge Works, exerting a discipline quite unlike that found elsewhere in the world. Greater Tokyo as well as Nagoya and Osaka/Kyoto/Kobe are among the most dynamic, difficult, and demanding high-tech markets in the world. Surviving in these markets requires flexibility, virtuosity, and unusual managerial effectiveness; above all, it requires listening and responding to customers.

In addition to Tokyo Effects, Knowledge Works emerged and evolved under four essential conditions:

1. Toshiba was and is a technology-driven corporation. Technology determines strategy, meaning technology shapes firm form and function. Technology is what Toshiba managers try to manage, and so technical considerations drive capital budgeting and strategy making, and not vice versa. In spite of the import of the main bank system and the volatility of the Tokyo Stock Exchange, science and technology, not finance, move Toshiba.

2. Toshiba plans for technical continuity and discontinuity in managing its people, products, and manufacturing processes. Because high-tech manufacturing involves the commitment of highly specialized resources that are perishable, often short-lived, and financially risky, Toshiba is prepared for any number of contingencies, including precipitous shifts in demand and, hence, the cancellation of projects that have already consumed a lot of resources. Knowledge Works, embodying a considerable yet coherent range of product/process capabilities, reduce risks of technical discontinuity while they leverage opportunities associated with technical continuity.

3. Toshiba's factories, like the company and all of its units, embody complex, social-economic governance systems. In Japan, governance is exercised primarily in the interest of employee, not shareholder, rights. Management in such circumstances requires a close and nearly constant management of people because people transfer, transform, and apply technology. Strategies that recognize this—the growing interdependence of human resources, technology, site- and relation-specific capabilities—carry the day in knowledge-intensive industries where employee rights are paramount.

4. Toshiba's emphases are on developing, renewing, and recycling corporate resources. Toshiba has moved beyond the come-from-behind, reverse engineering, technology acquisition processes of old and is dedicated today to companywide efforts to discover, acquire, and diffuse new knowledge and to reapply existing knowledge from whatever quarter, within and without the firm. An incessant emphasis on knowledge renewal and recycling culminates in large-scale organizational change strategies, like TP campaigning, that eliminate slack and redundancy and hopefully mobilize resources across the board.

All these conditions are important. Their cumulative expression during the past two or three decades has strongly impacted Toshiba's manufacturing organization and strategy. And the conditions that have framed Toshiba's development are paralleled by an exhaustive scrutiny of the evolution of the 200 largest industrial firms in Japan since 1918, and by a coevolution of factories, firms, and interfirm networks.[22] Their emergence has also been nourished by an overarching theoretical concept, namely *the economies of speed.*

In highly volatile markets, the fastest provider often wins the race both for lower costs and higher market prices because today's competition is not about selling goods and services of equal amount and quality. Traditionally, selling goods and services of relatively equal amount and quality was the competitive game, and as products and production processes became standardized, costs and prices both fell, usually to the advantage of late market entrants who benefited from the experience of others. Japan's industrial firms were among the best in the come-from-behind, late-to-market game.

But when markets, technologies, products and processes change rapidly, often unpredictably, first-to-market performance preempts fast-follower and fast-overtaker strategies. When speed is everything, products and manufacturing processes rarely reach a stage of high-volume production and standardization before there are basic changes to the dominant design and accompanying manufacturing/assembly processes. Fundamental and frequent changes in industry standards and dominant designs obviously work to the detriment of late market entrants. All else being equal, being first and fast to market are the winning strategies in time-based competition.

As speed has become central to industrial competitiveness, technology generation and transfer have become as important as technology discovery. As competition has become increasingly knowledge-intensive and technology-based, high-value-added output—not aggregate output, pure and simple—is an economy's most telling measure of strength and well-being. In sum, innovators take all in markets when economies of scale are negligible and economies of speed everything. Agility and nimbleness are not usually associated with firms of Toshiba's size (75,000 employees in Japan and 35,000 overseas) or complexity (six technology sectors, scores of divisions, and nearly 12,000 products). But Toshiba is a technology-driven firm with a strategy of delivering the best products to the widest markets in the shortest possible times. This has led to the creation of Knowledge Works, giving Toshiba the strength not only to stay in the race but to pull ahead of the competition as well.

The widespread existence of Knowledge Works in Japan but their relative scarcity (or nonexistence) elsewhere suggests the powerful impact of national competition on manufacturing organization at home, the competitive edge enjoyed by Japan's industrial firms in established product markets, and their ability to respond quickly, even preemptively, to new domestic and overseas markets. Manufacturing organizations of this sort also help explain Japan's prolonged improvement in manufacturing output per worker and steady increase in worldwide share of high-value-added products since the 1960s. Less obviously, the social embeddedness of such capabilities foreshadows the problems of taking this form of high-tech manufacturing organization abroad.

That is the message of site- and relation-specific skills, capabilities, and resources. But another message is that site- and relation-specific assets such as these may be nurtured, managed, and even recreated across institutional and national boundaries. Today Toshiba is successfully producing toner for photocopiers in the United States, toner and photocopiers in France, and photocopiers in China. So while imitation will not work, emulation, adaptation, and a lot of hard work will.[23]

Knowledge Works in Comparison

It is the integration and interpenetration of these circumstances, rather than their mere juxtaposition, that make all the difference in manufacturing strategy. Also, the ends and purposes of strategy make a big difference. Since private firms provide funds for about 80 percent of R & D and carry out 75 percent of it in Japan, new technologies are not cloaked in secrecy and readied for a military endgame.[24] Such considerations have long ruled against the commercialization of certain technologies in North America and Europe.

In other words, companies like Toshiba are *not* finance-driven and leading factories like Yanagicho are *not* regarded as simple, black-box production functions. On the basis of these conditions alone, one could argue for a number of different zones of manufacturing strategy worldwide. In North America, firms are more finance- or market-driven than technology-driven.[25] Performance is often assessed in the very short run according to financial criteria like net present value, quarterly profits, and returns on investment calculated according to product portfolio analysis techniques.

Since the pay-off period for big-ticket, technology-intensive projects is at least four to seven years down the road, by such financial criteria large industrial firms cannot afford to wait even three or four years to recoup their investments in *somewhat new* technologies. Certainly a 12-year cycle time for conceptualizing, refining, and implementing new and revolutionary technologies, like superconductivity or neural network computers, is beyond the kin of funding probability. In North America, it is quicker and easier to buy and sell companies (or divisions) as a means of obtaining new technical capabilities rather than developing and managing them oneself. New technologies tend to be shortchanged in such an environment, and as such they do not motivate and imbue corporate strategy. It is indeed sobering to realize how many global manufacturing firms are managed and measured by financial and not technical criteria.

In Europe, there is ample recognition of factories as complex social-economic systems even while there is no consensus with respect to which institutional arrangements maximize labor-management consensus and cooperation.[26] A separation of operational and strategic decision making in German firms between managerial and supervisory boards, for example, appeared as early as 1884 and the tradition lives on strongly today. But the impact of codetermination on performance, especially in global markets with mobile labor, capital, and technology, is unclear.

So far, the market for corporate acquisitions and takeovers has been less active in Europe than in North America, partly because there is greater recognition of the need to consider employee rights broadly rather than shareholder rights narrowly. But all too often recognition of employee interests in the form of profit and power sharing arrangements retard corporate performance because employment security, seniority, and job definition are too rigidly interpreted and guarded. Such circumstances make changes in work flows, job designs, new automation routines, and new staffing and skilling deployments, difficult if not impossible. The real success of *kaizen*, or continuous improvement, in Japan has been a willingness to experiment routinely and often with such circumstances.

It is principally in Japan (and perhaps in its East Asian neighbors of South Korea and the Republic of China as well as in the reunited Germany) that both conditions of industrial success are fulfilled: firms are technology driven *and* factories are recognized as complex, social-economic governance systems. Both conditions have everything to do with the timing of industrialization, the emergence of cooperative industrial relation systems, and a communal sense of shared risks and rewards in the process of industrialization.[27]

But it is not at all evident that management, industrial relations, and manufacturing strategy have jelled in South Korea, Taiwan, and Germany as they have in Japan. Douglass North argues that social institutions, such as property rights, legal codes, and systems of public education, provide a structure for economic exchange that along with technology largely determine transaction and production costs.[28] Class-like divisions still separate labor and management in South Korea, and rapidly increasing wages have dramatically altered production and R & D cost functions. Likewise, differences in the stock and quality of assets as well as in physical and social infrastructure spending between the former Western and Eastern Germanies suggest that they may have some distance to go before the conditions of economic progress, educational opportunity, and social equity that underpin Japan's postwar success are realized.

Manufacturing strategy in high-tech, highly turbulent environments requires a kind of factory sovereignty that is highly contextualized and that empowers employees in important organizational, managerial, and technical matters. In other words, intellectual capital is people-dependent capital. Hence, workplace-centered, cooperative human resource strategies are at the heart of Japan's industrial success, and several features of this approach are distinctive and seem related to timing:

1. Japan was the *last to industrialize* (beginning in the 1880s) among today's advanced industrial economies.
2. Further, Japan was the *last to reform its industrial structure and*

industrial relations system (sometime in the 1950s) among today's leading industrial nations.

After the Pacific War, both corporations and industrial relations were reworked and revised in light of the Allied Occupation (1945–51), and thus in view of particular American notions of industrial democracy, labor-management relations, and appropriate responsibilities for public corporations. For an average Japanese citizen during the 1950s and 1960s, public corporations were a primary institutional means, along with government, for recovering Japan's industrial might and for realizing new democratic ideals of individual participation and recognition in the workplace. The corporation was a crucible for national recovery and redefinition.

Firms found their way during the Korean War (1950–53) and ironed out kinks in a system of industrial relations and business-government relations by the end of the 1950s. Companies did this during an era of Western-inspired industrial democracy when labor unions were in ascendancy, the rights of workers enshrined, individualism and equality on the rise. They also did this with government support in the form of a developing communications, transportation, and educational infrastructure and in the context of emergent industrial policies aimed at labor-management accord, national recovery, economic restructuring, and social renewal. Nevertheless, companies accomplished the tasks of recovery, restructuring, and renewal largely on their own because day-to-day decisions were theirs to make, implement, abandon, or reinforce.

Toshiba's manufacturing strategy cannot be understood without first understanding these conditions and circumstances, their interrelations, and the context of postwar industrial relations that prefaced Japan's emergence as one the world's best performing industrial economies. The Cold War division of Germany and Korea may have replicated these circumstances somewhat, by linking corporate structures and industrial relations with advanced technology trajectories late in the 20th century. So the chronology driving Japan's industrial system of workplace-centeredness and knowledge-intensivity may not be unique, even if it is distinctive. Nevertheless, the drive to manage the structures, processes, and the meanings of modern factory work in Japan, as seen in Total Productivity campaigning and the governance of supplier relations, is remarkable and probably unprecedented.

How long Japan's industrial firms will sustain their outstanding performance and competitiveness into the next century will likely hinge on considerations similar to those of the last 30 to 40 years. Most importantly, these are the conditions whereby factories are fulcrums of intellectual capital and creativity, equitable industrial relations, effective supplier management, and motivated employees committed to organizational change and renewal. As long as Japan's firms retain these features, remain technology driven, and embody factories as complex, social-economic governance systems, that could be for a very long time.

Notes

Chapter 1

1. Interviews with persons responsible for manufacturing engineering practice at the Yanagicho Works were conducted over the 1987–90 period. These include Messrs. Shibamiya, Osada, Komai, Matsuda, Wada, and Domon, among others.

2. See such representative works by these authors as Jules Verne, *The Annotated Jules Verne: Twenty Thousands Leagues Under the Sea*, New York: Crowell, 1976; H. G. Wells, *The Time Machine*, New York: Heritage Press, 1964; Philip K. Dick, *Dr. Bloodmoney: or, How We Got Along After the Bomb*, Boston: Gregg Press, 1980, , and *Do Androids Dream of Electric Sheep*, London: Panther, 1972, ; and Ray Bradbury *Fahrenheit 451*, New York: Ballantine Books, 1953.

3. "Chingin kakumei: nenkoshugi to no ketsubetsu [Wage Revolution: The collapse of seniority-based compensation]," *Nihon Keizai Shimbun*, December 13, 1995, 1.

4. Lawrence G. Franko, "The Japanese Juggernaut Rolls On," *Sloan Management Review* 37 (2) (Winter 1996), 103–9.

5. *Nikkei Business*, "$1=80 Yen Factory," September 19, 1994 (No. 757), 10–22. Andrew Pollack, "Stunning Changes in Japan's Economy," *The New York Times*, October 23, 1994, section 3, 1.

6. David Teece, "Perspectives on Alfred Chandler's *Scale and Scope*," *Journal of Economic Literature* 31 (1) (March 1993), 199–225.

7. Peter Drucker, *Post-Capitalist Society*, New York: Harper Business, 1993; Robert Reich, *The Work of Nations*, New York, Vintage Books, 1992. Also see Keith Bradsher, "Skilled Workers Watch Their Jobs Migrate Overseas," *New York Times*, August 28, 1995, A1 and D6.

8. Karl Marx and Frederick Engels, *Manifesto of the Communist Party* (1955 trans.), Moscow: Foreign Languages Publishing House, 1955, 65.

9. See Hiroyuki Itami and Thomas W. Roehl's treatment of this theme in *Mobilizing Invisible Assets*, Cambridge: Harvard University Press, 1987.

10. C. K. Prahalad and Gary Hamel's "The Core Competence of the Corporation," *Harvard Business Review* (May-June 1990), 79–91, presents an argument that core competencies define what a firm is good at. I take the argument one step farther by suggesting that competencies organized and managed at the level of the factory are preeminent core competencies.

11. National Science Foundation, *The Science and Technology Resources of Japan: A Comparison with the United States*, NSF 88-318, Washington, D. C., 1988. See also, Economic Planning Agency, Economic White Paper, Section 2, Tokyo: Finance Ministry, 1990, 143–255.

12. Christopher Freeman, *Technology Policy and Economic Performance*, London: Pinters Publishing, 1987.

13. W. Mark Fruin and Toshihiro Nishiguchi, "Supplying the Toyota Production System—Intercorporate Organizational Evolution and Supplier Subsystems," in Bruce Kogut, ed., *Country Competitiveness and Organizing at Work*, New York: Oxford University Press, 1993, 225–246.

14. Martin Kenney and Richard Florida, "The Organization and Geography of Japanese R & D: Results from a Survey of Japanese Electronics and Biotechnology Firms," *Research Policy* 23 (1994), 305–23. Also, their *Beyond Mass Production*, New York: Oxford University Press, 1992, for a somewhat different formulation of the postproduction workplace.

15. Kazuo Sato, "Saving and Investment," in Kozo Yamamura and Yasukichi Yasuba, eds., *The Political Economy of Japan*, Vol. 1, *The Domestic Transformation*, Stanford: Stanford Univ Press, 1987.

16. M. Ishaq Nadiri, "Sectoral Productivity Slowdown," American Economic Review 70 (2) (1980), 349–52. Idem., "Contributions and Determinants of Research and Development Expenditures in the U.S. Manufacturing Industries," in G. M. von Furstenberg, ed., Capital, Efficiency, and Growth, Cambridge: Ballinger, 198.

17. Zvi Griliches, "R & D and the Productivity Slowdown," *American Economic Review* 70 (2) (1980), 343–48. Robert H. Hayes and Ranchandran Jaikumar, "Manufacturing's Crisis: New Technologies, Obsolete Organizations," *Harvard Business Review* (September-October 1988), 77–85. Robert E. Hoskisson and Michael A. Hitt, "Strategic Control Systems and Relative R&D Investment in Large Multiproduct Firms," *Strategic Management Journal* 9 (1988), 605–21.

18. Hiroyuki Odagiri, "Research Activity, Output Growth, and Productivity Increase in Japanese Manufacturing Industries," *Research Policy* 14 (1985), 117–30.

19. Koya Azumi and Frank Hull, "Inventive Payoff from R & D in Japanese Industry: Convergence with the West?" *IEEE Transactions on Engineering Management* 37 (1) (February 1990), 3–9; Frank M. Hull, Jerald Hage, and Koya Azumi, "R & D Management Strategies: America versus Japan," *IEEE Transactions on Engineering Management* 32 (2) (May 1985), 78–83; Frank M. Hull and Koya Azumi, "Teamwork in Japanese and U.S. Labs," *Research Technology Management* 32 (6) (November-December 1989), 21–26.

20. Edwin Mansfield, "Basic Research and Productivity Increase in Manufacturing," *American Economic Review* 70 (5) (1980), 863–73; Idem., "Competitiveness of American Manufacturing: Technological and Productivity Factors," *Managerial and Decision Economics* (Spring 1989), XXX–XXX; Idem., "The Speed and Cost of Industrial Innovation in Japan and the United States: External vs. Internal Technology, *Management Science* 34 (10) (October 1988), 1,157–68.

See also Albert N. Link, "Basic Research and Productivity Increase in

Manufacturing: Additional Evidence," *American Economic Review* 71 (5) (1981), 1,111–12.

21. A. Michael Spence, "The Learning Curve and Competition," *Bell Journal of Economics* 12 (1) (1981), 49–70; Hiroyuki Odagiri, "Research Activity, Output Growth, and Productivity Increase in Japanese Manufacturing Industries," *Research Policy* 14 (1985), 117–30; Hiroyuki Odagiri and Hitoshi Iwata, "The Impact of R&D on Productivity Increase in Japanese Manufacturing Companies," *Research Policy* 15 (1986), 13–19.

22. Ronald Dore, *British Factory-Japanese Factory*, Berkeley: University of California Press, 1973.

23. Robert E. Cole, *Work, Mobility, and Participation*, Berkeley: University of California Press, 1979. Idem., *Strategies for Learning*, Berkeley: University of California Press, 1989.

24. James R. Lincoln and Arne L. Kalleberg, *Culture, Control, and Commitment*, Cambridge: Cambridge University Press, 1990.

25. This definition of a Knowledge Works is not too different from the prescriptive definition for manufacturing offered by the Havard Business School. See, for example, Kim B. Clark, Robert H. Hayes, and Christopher Lorenz, *The Uneasy Alliance*, Boston: Harvard Business School, 1985; Robert H. Hayes, Steven C. Wheelwright, and Kim B. Clark, *Dynamic Manufacturing*, New York: Free Press, 1988. Otherwise, Masahiko Aoki explores some of the same issues in *Information, Incentives, and Bargaining in the Japanese Economy*, Cambridge: Cambridge University Press, 1988.

26. I deal extensively with issues of functional analogues in my book, *The Japanese Enteprise System—Competitive Strategies and Cooperative Structures*, Oxford: Clarendon Press, 1992 (also in paperback, 1994). When structure, strategy, process, and performance differ as notably as they do between Japanese and American corporations, the functional analogue approach is not an appropriate one.

Hugh Whittaker's book, *Managing Innovation* (Cambridge University Press, 1990), does compare Japanese and British factories along a number of important dimensions. However, the factories compared by Whittaker are all small factories, and size has an important effect on structure, strategy, and performance.

27. W. Mark Fruin, *The Japanese Enterprise System*, chapter 5.

28. These terms are defined more fully in chapter 1 of *The Japanese Enterprise System* cited above. For the moment, economies of scale signify decreasing costs per unit as a function of volume; economies of scope refer to decreasing per unit costs realized through a sharing of resources associated with the provision of multiple goods or services; learning economies are cost savings based on doing things more quickly with less effort, time, and material; transaction cost economies are savings on administrative or managerial overhead associated with linking functions, actors, and activities.

29. Fruin, *The Japanese Enterprise System*.

30. Recent studies of the overseas transfer of Japanese technology and organization include Tetsuo Abo, *The Hybrid Factory*, New York: Oxford University Press, 1992; Paul Adler, "The Learning Bureaucracy," in Barry M. Staw and Larry L. Cummings, eds., *Research in Organizational Behavior*, Greenwich, Conn: JAI Press, 1993; Mary Yoko Brannen, *Your Next Boss is Japanese:*

Negotiated Culture and Organizational Change, New York: Oxford University Press, in press; J. Fucini and S. Fucini, *Working for the Japanese: Inside Mazda's American Auto Plant*, New York: The Free Press, 1990; Martin Kenney and Richard Florida, *Beyond Mass Production*, New York: Oxford University Press, 1992.

31. According to Japanese government data, the industries with the highest intensity of R & D investment as a percent of sales are electrical equipment, chemicals (including pharmaceuticals), precision instruments, transportation equipment, steel, and general machinery. Given the intensity of R & D investment in these industries, organizational innovations to capture returns from R & D are more likely to occur here (Economic Planning Agency, *Economic White Paper*, Tokyo: Finance Ministry Press, 1990, 229–30). Also see Kikuchi Jun'ichi, Mori Shunsuke, Baba Yasunori, and Morino Yoshiki, *Kenkyu Kaihatsu no Dainamikkusu* [The Dynamics of Research and Development], National Institute of Science and Technology Policy (NISTEP), Science and Technology Agency, Report No. 14, September 1990.

32. There is a growing appreciation of the fact that productivity is not simply an economic problems but a social, political, and cultural one as well. See Mary Douglas, *How Institutions Think*, Syracuse: Syracuse University Press, 1986; Kasra Ferdows, ed., *Managing International Manufacturing*, Amsterdam: North Holland, 1989; Christopher Freeman, *Technology Policy, and Economic Performance*, London: Pinters, 1987; Paul Willman, *Technological Change, Collective Bargaining, and Industrial Efficiency*, Oxford: Clarendon Press, 1986.

33. The flexible specialization model is most closely associated with Michael Piore and Charles Sabel of the Massachusetts Institute of Technology, as argued in their *The Second Industrial Divide*, New York: Basic Books, 1984.

34. Both governments and corporations organize knowledge resources, and so information-to-knowledge transformation costs affect public and private interests. Yet, when governments build highways or even "information superhighways," apart from deciding basic licensing requirements and rules of the road, they do not decide who drives on them, how well they drive, or where they go. Individuals and firms, private actors for the most part, do.

35. Wolhee Choe, *Towards an Aesthetic Criticism of Technology*, New York and Bern: Peter Lang, 1989, 5–6. Also Franz Boas, *Race Language, and Culture*, Chicago: University of Chicago Press edition, 1982, 562–63, 588–92.

36. I have adapted this figure, changing how the angles and axes are named, from one developed originally by Dr. Venky Narayanamurti, dean, College of Engineering, University of California, Santa Barbara.

Chapter 2

1. Simon Kuznets, *Modern Economic Growth: Rate, Structure, Spread*, New Haven: Yale University Press, 1966, p 6, but see discussion on pp 6–15.

2. W. Mark Fruin, *The Japanese Enterprise System—Competitive Strategies and Cooperative Structures*, Oxford University Press, 1992, chapters 3 and 4.

3. Alfred D. Chandler, Jr., *Scale and Scope*, Cambridge: Harvard University Press, 1990.

4. There is a vast literature on quantifying optimum scale for different factories, based on industry, market, and other macroeconomic factors. There is also a counterliterature, such as the treatise *Limits of Organization*, (New York: W. W. Norton, 1974, by Kenneth Arrow. Arrow argues that beyond a certain size or scale, diseconomies of scale appear. The point being that economies and diseconomies of scale are not independent of a host of organizational, technical, and strategic factors that are nearly impossible to model mathematically.

5. James Abbegglen and George Stalk, Kaisha—The Japanese Corporation, New York: Basic Books, 1985, 120–22.

6. When R & D spending declines, as it did by 16.7 percent for Toshiba in the 1992–93 fiscal year, the impact is felt more keenly in R & D facilities that are geographically and organizationally removed from multifunction, multiproduct production sites. Obviously, a Knowledge Works' R & D budget can be cut as well, but the greater integration of research and design activities in production at Knowledge Works insulates them to a degree from annual budget fluctuations. *Nihon Keizai Shimbun,* "Toshiba, 16.7% Shimogata Shusei," September 1, 1992, 1.

7. Contrast these 3 K's with the more popularly voiced 3K's of manufacturing heard frequently in Japan during the late 1980s, *kitsui, kiken,* and *kitanai,* or difficult, dangerous, and dirty. But the popular characterization arose in contrast to the allure of go-go, fast-track, high-paying careers in Kayabacho, or the Tokyo equivalent of Wall Street. With the worldwide collapse of financial markets in the late 1980s and early 90s, fast-track careers in finance and real estate began to look more *kitsui, kiken,* and *kitanai* than knowledge-intensive engineering and sales. A fourth K, one especially attuned to the product and process innovation activities of a Knowledge Works, might be *kokoro,* a Japanese word meaning spirit, heart, soul.

8. There have been a number of studies that identify factors that are important to improving factory productivity, although in general, they do not take up issues of how multifunction and multiproduct capabilities should be organized in single sites and how these capabilities may affect productivity. See, for example, Robert Hayes and Kim B. Clark, "Why Some Factories Are More Productive than Others," *Harvard Business Review* (September/October 1986), 66–73; Roger W. Schmenner, "Behind Labor Productivity Gains in the Factory," *Journal of Manufacturing and Operations Management* 1(4) (Winter 1988), 323–38; Roger W. Schmenner and Boo Ho Rho, "An International Comparison of Factory Productivity," *International Journal of Operations and Production Management* 5 (3) (May 1985), 273–89.

9. The concept of relation-specific skill is Banri Asanuma's. See his "Manufacturer-Supplier Relationships in Japan and the Concept of Relation-Specific Skill," *Journal of the Japanese and International Economics* 3 (1989), 1–30; also Asanuma Banri, "Setsubi toshi kettei no prosesu to kijun (2)," *Kyoto Daigaku Keizai Ronso,* 130 (5/6) (November/December 1982), 23–44.

10. Fruin, The Japanese Enterprise System, chapters 4 and 5 present de-

tailed information on the organizational development of Hitachi, Ltd. and Matsushita Electric Industrial.

11. Toshiba Corporation, Toshiba factory brochures, 1990–92. Place of publication and publisher not given.

12. One of the more popular interpretations of Japanese manufacturing prowess emphasizes an overlapping of functional requirements in single-product factories. No doubt overlapping occurs and it enhances performance. However, I question the ease with which many scholars assume that overlapping occurs and I doubt that there are many large, single-product factories in Japan. In short, if we want to understand what is happening in Japanese manufacturing today, we have to get out of the classroom and onto the shop floor. See, Ken'ichi Imai; Ikujiro Nonaka; and Hirotake Takeuchi, "Managing the New Product Development Process: How Japanese Companies Learn and Unlearn," in Kim B. Clark, Robert H. Hayes, and Christopher Lorenz, eds., *The Uneasy Alliance*, Boston: Harvard Business School, 1985.

There have been only three books published in English about manufacturing in Japan with a field work emphasis, that is, an emphasis that takes factories as whole and distinct units of analysis and interpretation. Two of these looked at smaller-sized factories and only one took up large-scale factories, comparable to the Yanagicho Works. None of these works looked at the history of factories as an independent variable and none studied complex factories that could be considered analogous to Knowledge Works.

The first of these, James Abegglen's *The Japanese Factory* (Free Press, 1958) was long a classic, not only in Japanese Studies but also more generally in social science literature. Abegglen's thesis positing a link between the Japanese factory and the culture and history of Japan has not been sustained by subsequent scholarship either in Japan or in the West. Robert Cole's *Japanese Blue Collar* (University of California, 1971) offered a straightforward structural-functional explanation for the Japanese workplace. Cole argued that whatever differences characterized Japanese factories as opposed to Western factories were not substantial, and they were largely attributable to differences in the pace of economic development.

Finally, Ronald Dore's *British Factory-Japanese Factory* (University of California Press), which appeared in 1973, compared the General Electric Company of Great Britain with Japan's Hitachi. The comparative framework highlighted actual differences in function and structure that distinguished GEC and Hitachi. Dore's interpretations of these differences emphasized an organization-centered society (Japan) as opposed to an individual-centered one (the United Kingdom), plus hypotheses of technological convergence and late development.

13. Robert H. Hayes, Steven C. Wheelwright, and Kim B. Clark, *Dynamic Manufacturing—Creating the Learning Organization*, New York: The Free Press, 1988, 275–78.

14. Jane Jacobs, *Cities and the Wealth of Nations: Principles of Economic Life*, New York: Random House, 1984. See also the new history of New York, the preeminent American city; Kenneth T. Jackson, *The Encyclopedia of New York City*, New Haven and New York: Yale University Press and the New York Historical Society, 1995.

15. In rank order, these are Tokyo-Yokohama, Mexico City, New York, Sao Paulo, Seoul and Osaka-Kyoto-Kobe. "Unrest: Urban Turmoil Throughout the World," *Los Angeles Times*, Monday, May 25, 1992, A1, 36–37.

16. The model of fast-to-market competition presented here is very different from other models of industry competition in Japan. For example, Kozo Yamamura argues that steel investment cartels, operating under government policies that awarded output shares on the basis of market share and capital investment, created a system of overinvestment and excess capacity. See "Success that Soured: Administrative Guidance and Cartels in Japan," in Kozo Yamamura, ed., *Policy and Trade Issues of the Japanese Economy*, Seattle: University of Washington Press, 1982, 77–112. However, information technology products and services markets are very different from steel markets, and there is very little evidence that information technology companies operate as cartels or that government policy making has had the same effect across all industries. See Martin Fransman's book, *Japan's Communications and Computer Industry*, Oxford: Oxford University Press, 1995. Also, on hypercompetition, see Richard A. D'Aveni with Robert Gunther, *Hyper-competitive Rivalries*, New York: The Free Press, 1994.

17. Chikashi Moriguchi, *Nihon Keizairon*, Tokyo: Sobunkan, 1988, 15–17.

18. For a discussion of high-tech and high-touch features of modern work, see Michael Maccoby's "Forces for Increased Organizational Competition: Implications for Leadership and Organizational Competence," in Daniel B. Fishman and Cary Cherniss, eds., *The Human Side of Corporate Competitiveness*, Newbury Park and London: Sage Publications, 1990, 35–50.

19. The point of how rapidly firms have to invest to stay even in times of rapid economic growth is made in James C. Abegglen and George Stalk, Jr., *Kaisha, The Japanese Corporation*, Basic Books, 1985. At the same time that Japan's firms were investing at unprecedented levels in upgrading existing capabilities and in developing new ones, American firms were in the midst of the biggest buying and selling frenzy of company assets in modern times.

The response of leading firms in the two economies could not have been more different. The American federal deficit was increasingly financed by the sale of U.S. treasury bonds to Japanese buyers, while at the same time a number of American firms with sizeable real estate holdings in Japan chose to downsize or liquidate their assets there in order to take advantage of the runup in property values in Tokyo. For a discussion of the merger mania in America, see James B. Stewart, *Den of Thieves*, New York: Simon & Schuster, 1991.

20. Moriguchi, *Nihon Keizairon*, 4–6, 15.

21. The history section of this chapter was translated and adapted from *Toshiba Yanagicho Yonjunen no Ayumi*, Kawasaki City, 1976, and *Sozo - Toshiba Yanagicho Kojo Gojunenshi*, Kawasaki City, 1987. The details of product development and organizational adaptation, as outlined here, provide basic evidence to support the hypothesis that organizations can learn and grow through experience and intention.

22. Mary Douglas, *How Institutions Think*, Syracuse University Press, 1986.

23. Wickham Skinner, "The Focused Factory," *Harvard Business Review* (May-June 1974), 113–21; Stefan Aguren and Jan Edgren, *New Factories*, Stockholm: Swedish Employers' Confederation, 1980; Roy L. Harmon and Leroy D. Peterson, *Reinventing The Factory*, New York: Basic Books, 1990. A discussion of general issues in production organization can be found in Kim B. Clark, Robert H. Hayes, and Christopher Lorenz, eds., *The Uneasy Alliance: Managing the Productivity-Technology Dilemma*, Harvard Business School Press, 1985.

24. For a longer discussion of the importance of well-defined, internal organizational boundaries, see W. Mark Fruin, "Good Fences Make Good Neighbors—Organizational Property Rights and Permeabiity in Product Development Strategies in Japan," paper delivered at the Workshop on Competitive Product Development, Euro-Asia Centre, INSEAD, June 27–29, 1991, Fontainebleau, France, 38 pp.

25. There are at least 18 different kinds of routine planning meetings occurring within the Information and Communications Group at Yanagicho. As the frequency of these meetings hinges on the degree of product and process complexity and market pull for Yanagicho's information products lineup, many of the meetings occur on a weekly and occasionally a daily basis. A count of factory telephone and fax connections was taken from a 1991 directory.

26. Interview with Dr. Hiroshi Komiya, general manager, Mitsubishi Electric Corporation, LSI Research and Development Laboratory, Itami City, Japan, February 8, 1991.

27. Kenichi Ohmae, *The Borderless World*, New York: Harper Business, 1990.

28. *Wall Street Journal*, "Long Road Ahead: Detroit Needs Big Overhaul to Match Japanese," January 10, 1992, A1, 6.

29. *Los Angeles Times*, "United Technologies to Cut 13,900 Jobs in Restructuring," January 22, 1992, D1–2.; telephone interview with Mr. John Sandberg, United Technologies corporate staff, Jan. 22, 1992.

30. Paul Burnham Finney, "Executives on the Go Are the Main Buyers of Portable Computer," *The New York Times*, January 17, 1996, C3.

31. Nomura Research Institute, *TOSHIBA NRI Report*, Tokyo, February 5, 1990, 52 pp.

32. Interview with Mr. Y. Iwatsuki, Planning Department, Yanagicho Works, Kawasaki City, August 4, 1987.

33. Data are taken from the Yanagicho Works Statistical Handbook, updated and photocopied quarterly, June 1989, 7.

34. Interview with Mr. Tetsuya Yamamoto, senior manager, Logistics, Toshiba Corporation, Shibaura, Tokyo, April 3, 1992.

35. In ancient Greek, *prometheus* means forethought. In Greek mythology, Prometheus, a Titan, stole food from Zeus and fire from Olympus as a champion of men against the Gods. The image of Prometheus as a provider of fuel and fire for humankind's benefit inspires this study of high-tech manufacturing strategy.

36. Nomura Research Institute, *NRI Report*, 35.

37. Kim B. Clark and Takahiro Fujimoto, *Product Development Performance*, Boston: Harvard Business School Press, 1991. The new Boeing 777 jet-

liner was built using "design-build teams," a concept picked up in Japan, probably looking at the product development organizations described by Clark and Fujimoto. For a description of Boeing's design-build teams, see Karl Sabbagh, *21st-Century Jet*, New York, Scribner, 1996.

38. The extreme emphasis during the 1980s on first-to-market production of many different products and models has diminished somewhat during the 1990s. A MITI study cited by *The New York Times* found that during the 1991 FY, 211 different models of television were offered, remaining on the market an average of 12 to 13 months. The projected figures for 1992 were 192 different models and a 14-month cycle time. A move toward somewhat fewer models and slightly longer cycle times was found across the consumer electronic and auto industries. Andrew Pollack, "Japan Eases 57 Varieties Marketing," *The New York Times*, October 15, 1992, D1, D7.

Nevertheless, Knowledge Works, by their nature and purpose, cannot be isolated from Tokyo Effects, a concentration of highly skilled and qualified people, fashion-conscious markets, and discrimating consumers. Even so, Tokyo Effects are neither confined to Tokyo, nor even to Japan. MacMillan Bloedel Limited, Canada's largest forest resources company, designs, cuts, and packages lumber housing kits for architects working in fast-growing markets south of the border. Finished housing runs $150 per square foot in Palo Alto, California, for example, but only half as much in greater Vancouver. So, Palo Alto architects draw blueprints for Silicon Valley clients, have MacMillan Bloedel design, cut, and package the lumber, and hire framing specialists to assemble the packages on local lots. A classic Tokyo Effect. Interview with Roger N. Wiewel, senior vice president for product development, MacMillan Bloedel, January 30, 1991, Tokyo.

Chapter 3

1. Integrity, as it is used here, refers to the coherence and unity of organizational strategy, structure, and process, as used by Kim Clark and Takahiro Fujimoto, *Product Development Effectiveness*, Boston: Harvard Business School Press, 1992. Also, Julian Gresser is increasingly using integrity as a key concept in his work on effective action and negotiating with the Japanese; see his *Piloting Through Chaos*, Sausalito, Calif.: Five Rings Press, 1995.

2. According to a 1993 survey of 16,300 business organizations with a paid-in capital of 5 million yen or more conducted by the Management and Coordination Agency, for the first time royalty receipts exceeded royalty payments for technology trade. Of course, for large, technology-intensive firms like Toshiba, royalty receipts have exceeded royalty payment for quite some time, in some cases since the oil shocks of the 1970s. *The Daily Yomiuri*, "Technology Trade Runs Surplus in Fiscal 1993," January 5, 1995, 12.

3. Harvey Leibenstein, "Aspects of the X-efficiency Theory of the Firm," *Bell Journal of Economics* 6(2) (Fall 1975), 580–606. Idem., *Beyond Economic Man*, Cambridge: Harvard University Press, 1976. For a more recent treatment of adaptation, see Kathleen M. Eisenhardt and Behnam N. Tabrizi, "Accelerating Adaptive Processes: Product Innovation in the Global

Computer Industry," *Administrative Science Quarterly*, 40(1) (March 1995), 84–110.

4. Paul J. DiMaggio and Walter W. Powell, "The Iron Cage Revisited: Institutional Isomorphism and Collective Rationality in Organizational Fields," *American Sociological Review* 48 (April 1983), 147–60.

5. Michel Crozier, *The Bureaucratic Phenomenon*, London: Tavistock Institute, 1964. James Abegglen, *The Japanese Factory*, Glencoe, Ill.: The Free Press, 1958.

6. Etzioni's four phases are similar to the three phases of TQM discussed by one of NEC's top design engineers, namely, awareness, empowerment, and alignment. See Kiyoshi Uchimaru et. al., *TQM for Technical Groups*, Portland, Ore.: Productivity Press, 1993. Amitai Etzioni, *Political Unification: A Comparative Study of Leaders and Forces*, New York: Holt, Rinehart and Winston, 1965.

7. Joseph Juran, "Why Quality in Japan," *Washington Post*, op/ed commentary page, August 15, 1993.

8. Japan Management Association, *TP Manejimento '91 Nendo*, Tokyo: JMA, 1991, 6–7.

9. *Toshiba Newsletter*, February 1988, No. 307, 1.

10. Interview with Tetsuya Yamamoto, Tokyo, August 7, 1991.

11. Ibid.

12. Sumiyoshi interview, Yanagicho, 2/25/91. The key points of TP management, as seen by Yamamoto, are: 1) to set clear goals for the entire organization; 2) to provide mechanisms whereby every department and section reviews the goals in light of their own means and ends; 3) to create the tools and standards of comparison for setting appropriate targets and rewards; 4) to train leaders for TP campaigning in local units and provide them with structures for integrating employee efforts.

13. Rodosogishi Kenkyukai, ed., *Nihon no Rodo Sogi 1945–80* [Japan's Labor Strife 1945–80], Tokyo: Tokyo University Press, 1991, 423–30.

14. Ibid., 439–52.

15. Interviews with Yasuo Kanesaki were recorded during the summer of 1988 by myself and Christena Turner, an anthropologist, who was visiting the Yanagicho Works at my invitation and helping analyze issues of organizational structure and culture. While our interviews with Kanesaki were conducted independently, the content of our interviews are merged in this section.

16. The assertion that Yanagicho employees, even the white collar ones, spend most of their careers at Yanagicho runs counter to most of the career development literature on Japanese employment practices. The typical story is that Japanese firms practice widespread job rotation and aim to create a management cadre of functional generalists. While I have no reason to doubt that this may be true of corporate managers, I have plenty of interview evidence that this is not true of factory-based employees. Consider two facts: 80 percent of Toshiba employees are located in factories, and that factories operate and advance on the basis of site- and relation-specific knowledge, not general knowledge. In other words, large industrial firms like Toshiba have two career cultures, one for functional generalists at the corporate level, and one for functional specialists at the factory level of operations.

17. Edgar H. Schein, "Organizational Culture," MIT Working Paper #2088-88, December 1988, 7. For a full exposition of Schein's thinking on this and other points, see *Organizational Culture and Leadership*, San Francisco: Jossey Bass, 1985.

18. Interview with Messrs. Shiba and Domon, Irvine, California, 1/26/90.

19. Interview with Mr. Chiba, Yanagicho, August 16, 1988.

20. Two examples of the negotiation of work culture in a Japanese-American and Japanese context are Mary Yoko Brannen, *Your Next Boss is Japanese*, University of Massachusetts, Amhert, Department of Anthropology and School of Management joint Dissertation, 1994; and Christena Turner, *Breaking the Silence*, Stanford University, Department of Anthropology Ph.D. Dissertation, 1988.

21. Researchers miss this important point because either they are too acclimated to classical Western notions of organizational control and managerial authority or they have spent too little time on the shop floors of large, industrial firms in Japan.

22. Most of this section is based on an interview with Mr. Sato, head of the Machine Tool Department at Yanagicho on August 16, 1988. Sato entered the Department in 1943 and was the department head (*kacho*) in 1988, 45 years later.

23. Mr. Sato, department head since 1970, whom I interviewed in the summer 1988, pointed out that the very first head of the department, Sato Takaya, was a graduate of Tokyo University who later became a managing director (*senmu torishimariyaku*) of Toshiba. Sato interview, Yanagicho Works, August 16, 1988.

24. For a complete description of the decentralization of key staff functions to product departments, see my discussion in *The Japanese Enterprise System: Competitive Strategies and Cooperative Structures*, Oxford: Oxford University Press, 229–36.

25. TP campaigning results for the Machine Tool and Labor Savings Equipment Departments are contained in various internal documents prepared by department managers for TP campaign managers. Since the documents were prepared on-site for the purposes of local evaluation and record keeping, I have no reason to believe that they are anything but accurate.

26. Yanagicho Works, "3C-WG" Action Plan, mimeograph, September–October 1990, 4 pages.

27. Interview with Mr. Namikawa, August 25, 1988, Yanagicho Works, Kawasaki City.

28. Yanagicho Works, "62nen/kami Kojo Undo Katsudo Hokoku," memograph, undated, 1 page.

29. Yanagicho Works, *Heisei Nisannen TP Katsudo no Hoshin*, photocopied, undated.

30. Tetsuya Yamamoto, Hammatsucho interview, August 1991.

31. Sumiyoshi interview, Yanagicho, February 25, 1991.

32. Okajima interview, Kawasaki City, Yanagicho Works, August 18, 1988; Yamamoto Tetsuya interview, August 26, 1993.

33. S. Shane, S. Venkataraman, and I. MacMillan, "The Effects of Cultural Differences on New Technology Championing Behavior within Firms," *Journal of High Technology Management Research* 5(2) (Fall 1994), 163–82.

34. Chiba interview, Yanagicho, August 16, 1988. Recent American literature points to renewed emphasis on human resource management as American businesses struggle to compete with the Japanese. As one American scholar has put it, "People seem to matter in direct proportion to an awareness of corporate crisis." Rosabeth Moss Kanter, *The Change Masters*, New York: Simon & Schuster, 1983, 17.

35. Interview with Yamamoto Tetsuya, Tokyo, August 26, 1993.

36. Interviews with Mr. Kondo, head, Training Department, Yanagicho Works, Spring 1991. The training categories are: general training, skills training, technology training, production management training, quality assurance training, sales engineering and service training, high school equivalency training, and company socialization programs.

37. Toshiba Corporation, *Personnel Management*, Tokyo: mimeo, undated but prepared in 1986 or 1987, 12 pp, with three appendices.

38. Interview with Mr. Sumiyoshi, Yanagicho, 2/25/91.

39. Interview with Yamamoto Tetsuya, Toshiba Corporation, Shibaura, Tokyo, August 26, 1993.

Chapter 4

A shorter version of this chapter appears in *Networks and Markets: Pacific Rim Strategies*, W. Mark Fruin, ed., Oxford University Press, forthcoming 1997–98.

1. These figures are based on what Toshiba calls the turn-out value (TOV), or internal transfer price value, of the factory's output to various divisions in Toshiba. The TOV is a concept that Toshiba borrowed from the General Electric Corporation in the 1970s. The figures themselves come from Yanagicho's Purchasing Department, *Supplier Management Guidelines*, mimeo, annual editions. These figures are from the 1988 edition.

2. Interview with Mr. Yoshii of the Purchasing Department, Yanagicho Works, Kawasaki City, August 9, 1988.

3. Hajime Oniki, "On the Cost of Deintegrating Information Networks," in C. Antonelli, ed. *The Economics of Information Networks*, Amsterdam: Elsevier Science Publishers, 1992, 397–410.

4. W. Mark Fruin, "Competing the Old-Fashioned Way: Localizing and Integrating Knowledge Resources in Fast-to-Market Competition," in J. Liker, J. Ettlie, and J. Campbell, eds., *Engineered in Japan: Technology Management Practices in Japan*, New York: Oxford University Press, 1995, in press.

5. W. Mark Fruin and Toshihiro Nishiguchi, "Supplying the Toyota Production System: Intercorporate Organizational Evolution and Supplier Subsystems," in Bruce Kogut, ed., *Country Competitiveness*, New York: Oxford University Press, 1993, 225–49. Nishiguchi's research on suppliers may be found in *Strategic Industrial Sourcing*, New York: Oxford University Press, 1994.

Two studies that look at interunit interdependence or networking and use it to predict how governance is organized to achieve integration are Andrew H. Van de Ven and G. Walker, "The Dynamics of Interorganizational Coordination," *Administrative Science Quarterly* 29(4), 598–621; C. Oliver,

"Determinants on Interorganizational Relationships: Integration and Future Directions," *Academy of Management Review* 15(2), 241–65.

6. Personal communications, Purchasing Department, Yanagicho, 1989, and the *Supplier Management Guidelines*.

7. Banri Asanuma, "Transactional Structure of Parts Supply in the Japanese Automobile and Electric Machinery Industries: A Comparative Analysis," *Technical Report No. 1*, Socio-Economic Systems Research Project, Kyoto: Kyoto University, July 1985.

8. Kim Clark and Takahiro Fujimoto, *Product Development Performance*, Boston: Harvard Business School Press, 1991.

9. Banri Asanuma, "Yasashi Keizaigaku," Nihon Keizai Shimbun, February 21–25, 1984. Idem., "The Structure of Parts Transactions in the Automotive Industry—Mechanisms of Adjustment and Innovative Adaptation," *Gendai Keizai* 59 (Summer 1984).

10. W. Mark Fruin, "Focal Factories," chapter 6, *The Japanese Enterprise System*, Oxford: Oxford University Press, 1992, for more details. Masahiko Aoki also emphasizes the coordinating role of factories in corporate strategy. See his "Horizontal vs. Vertical Information Structure of the Firm," *The American Economic Review* 76(5) (Dec. 1986), 971–83.

11. Personal communication, Tetsuya Yamamoto, senior manager, Logistics, Toshiba Corporation, Shibaura, Tokyo, April 3, 1992.

12. Most of the details on Yanagicho's relations with suppliers comes from the Purchasing Department of the Yanagicho Works where I conducted field interviews from 1986 through 1989 inclusive.

13. Fruin, *The Japanese Enterprise System*, chapter 1, 20–21.

14. Michael Gerlach, *Alliance Capitalism*, Berkeley: University of California Press, 1992. Gerlach provides the most up-to-date discussion of *keiretsu* groups in English.

15. Personal communication, Purchasing Department, Yanagicho Works, Summer 1987; Banri Asanuma, "Transactional Structure of Parts Supply," 1985.

16. Personal communication, Purchasing Department Section Manager, Yanagicho Works, Summer 1987.

17. A longer treatment of the restructuring issue can be found in Fruin, *The Japanese Enterprise System*, 1992, 243–47.

18. Again, details concerning the Yanagicho Works and its suppliers were supplied by the Purchasing Department of the factory during a period of several years when I was involved with fieldwork interviews and participant observation at the factory.

19. Yanagicho Purchasing Department, *Supplier's Management Guidelines*, mimeo, 1987, 4, for internal circulation only.

20. The information contained in this section comes from Nomura Research Institute, TOSHIBA, *NRI Report*, Feb. 5, 1990, 52 pp.

21. See *The Japanese Enterprise System*, 239–43, for a detailed description of the development of photocopiers and automatic mail sorting equipment at the Yanagicho Works.

22. For a discussion of this point, see Seiichi Kawasaki and John McMillan, "The Design of Contracts: Evidence from Japanese Subcontracting," *The Journal of Japanese and International Economies* 1(3) (1987), 327–49.

23. W. Mark Fruin, *The Japanese Enterprise System*, 239–43.

24. Nomura Research Institute, Toshiba, *NRI Report*, February 5, 1990, 38–46.

25. For example, the basic research on the third-generation, 64-mega DRAM chip technology was done at Toshiba's ULSI labs, and then transferred to SSDC for further development and refinement in 1990. Already, a clear division of labor in R & D activities is evident between Central R & D Labs, ULSI, and SSDC efforts. Three years earlier, when the ULSI Labs were perfecting the 16-mega DRAM technology, 4-mega DRAM technology had been passed to SSDCs for further development before sending the technology along to the Oita Works for final prototyping and volume manufacturing.

26. Walter W. Powell, "Neither Market nor Hierarchy: Network Forms of Organization," *Research in Organizational Behavior* 12 (1990), 295–336.

27. Permeability is a concept developed in my 1992 book, *The Japanese Enterprise System*, and in a 1995 paper, "Good Fences Make Good Neighbors."

28. These figures are based on a 1988 survey of members in the factory Supplier Association as conducted by the Purchasing Department of the Yanagicho Works.

29. In *The Japanese Enterprise System*, I referred to this competitive-cooperative dynamic as "competitive strategies and cooperative structures," the subtitle of my 1992 book.

30. Paul Sheard, "Japanese Corporate Boards and the Role of Bank Directors," in W. Mark Fruin, ed., in *Networks and Markets: Pacific Rim Strategies*, in press; another excellent and general article on *keiretsu* is James R. Lincoln, Michael L. Gerlach, and Christina Ahmadjian, "*Keiretsu* Networks and Corporate Performance in Japan," forthcoming in *American Sociological Review*.

31. Yanagicho Purchasing Department, *Suppliers Management Guidelines*, 1988 edition, 4.

32. Given how ordinary interorganizational transactions are, until one defines the nature of the transactions between firms and the intermediating mechanisms by which firms interact, it is impossible to say anything conclusive concerning interorganizational coordination, aside from the frequency of interorganizational transactions.

33. It has been suggested that Japanese firms use subcontracting or outsourcing rather than vertical integration as ways of internalizing market transactions. It is less an either/or choice than a question of emphasis. Also, such an observation appears more appropriate for the automobile industry than for industrial electronics where outsourcing is more strongly related to a full-line product strategy and less strongly related to time-to-market issues. However, the real point is that Japanese manufacturers have a range of organizational choices with respect to how they mobilize resources for competition.

Chapter 5

1. SuperSmart is a trademark name registered in Japan by the Toshiba Corporation.

2. Lewis Mandell, *The Credit Card Industry*, Boston: Twayne Publishers, 1990, xi.

3. Ibid., 150.

4. Toshiba IC Card Department, internal data, 1990.

5. Mandell, *Credit Card*, 151–52.

6. A detailed history of the product development accomplishments of the Yanagicho Works is contained in W. Mark Fruin, *The Japanese Enterprise System—Competitive Strategies and Cooperative Structures*, Oxford University Press, 1992.

7. See chapter 4 for a lengthy discussion of supplier relations when conventional parts, components, and subassemblies are being outsourced.

8. Compare these stages of development with those outlined for Bayer A. G., one of the first research-intensive corporations in the West. See Margaret Graham and Betty Pruett, *R & D for Industry*, Cambridge: Cambridge University Press, 1991.

9. Toshiba Corporation, Yanagicho Works, *Reports to Visa International*, memo, 1986–89, San Mateo, California, September 1, 1988 meeting, desktop published. Kubo's corporate championing of a SuperSmart card was clearly more important than Matsuda's at the factory level of organization. This is not to say that Matsuda's, and subsequently Tamada's, project leadership was unimportant, only that the championing of the SuperSmart card was something that required high-level legitimacy and access to the highest levels of decision making. In this sense, the SuperSmart card story supports Waldman and Atwater's view of the effectiveness of championing behavior. David A. Waldman and Leanne E. Atwater, "The Nature of Effective Leadership and Championing Processes at Different Levels in a R & D Hierarchy," *The Journal of High Technology Management Research* 5(2) (Fall 1994), 233–46.

10. Toshiba Corporation, *Reports to Visa*, report of October 20, 1986 meeting.

11. Interview Mr. Masuo Tamada, the Yanagicho Works, 2/20/90.

12. Ibid. Also, J. B. Quinn, H. Mintzberg, and R. M. James, *The Strategy Process*, New York: Prentice Hall, 1988, 700–02.

13. Interview with Mr. Hiro Shogase, Toshiba Headquarters, Shibaura, August 29, 1993.

14. Interview with Masuo Tamada, Yanagicho Works, April 4, 1987.

15. Ibid.

16. The meetings were held between October 25, 1986, and March 13, 1987. Toshiba Corporation, *Senryaku kaigi jisshi keiei oyobi yoyakuhyo 61/shimo* [Strategic Planning Committee Meetings and Schedule for the latter half of the 1986], single sheet, undated, presumably 1986.

17. Visa International, internal memo, August 7, 1987.

18. Ibid.

19. Toshiba Corporation, Yanagicho Works, *Reports to Visa International*, April 24, 1986, desktop published.

20. Toshiba Corporation, Yanagicho Works, *Reports to Visa International*, August 7, 1987, desktop published.

21. Toshiba Corporation, Yanagicho Works, *Reports to Visa International*, May 28, 1987, desktop published.

22. Visa International, Progress Report—Key Component Improvements, item 6.1, September 24, 1987.

23. Ibid.

24. Toshiba Corporation, Yanagicho Works, *Reports to Visa International*, September 24, 1987, desktop published.

25. March 2, 1988 progress report found at Visa International, San Mateo, California. Presumably this progress report was authored by Yanagicho engineers.

26. Memo from Bill Chen, sent by Gretchen McCoy, to Kouji Kuramochi, March 22, 1989; found in Visa International's San Mateo California offices.

27. A slightly different set of estimates were prepared four months earlier, in March 1988. The main differences were found in the length of the forward projections for unit prices and volume. The earlier estimates extended until 1993, two years beyond the September projections. What a difference several months make! In March, Toshiba was projecting sales of 80 million cards at a unit cost of $7.79 by 1993. The September figures were much more modest: 5 million cards in 1991 at a unit cost of $9.27. The September projections did not extend to 1992 and 1993. Toshiba Corporation, Yanagicho Works, *Progress Reports to Visa International*, company documents, dated March 1 and June 21, 1988, found at Visa International in San Mateo, California.

28. EEPROM technology, an area where Toshiba was relatively weak previously, advanced considerably as a result of Toshiba's development work on the SuperSmart card.

29. Toshiba Corporation, Yanagicho Works, *Reports to Visa International*, January 31, 1989 and May 11, 1989, desktop published.

30. Letter from Ms. Gretchen McCoy to Mr. Mitsuo Kubo, dated January 11, 1989 and found at the San Mateo, California offices of Visa International.

31. Letter from Mr. Charles Russell to Mr. Mitsuo Kubo found at the San Mateo California offices of Visa International.

32. The point that people often move with technology down the value chain in the Japanese system of R & D has been made before. However, earlier research made the point for connecting basic and applied R & D, while this research makes the connection between applied R & D and product design and development. See Kiyonori Sakakibara and D. Eleanor Westney, "A Comparative Study of the Training, Careers, and Organization of Engineers in the Computer Industry in the United States and Japan," *Hitotsubashi Journal of Commerce and Management* 20(1) (December 1985), 1–20.

33. Anthony L. Velocci, "Boeing, Labor Talks Ripple Industrywide," *Aviation Week & Space Technology*, August 14, 1995, 20–21.

34. Interview with Mr. Yoshinaga, Yanagicho Works, Kawasaki City, August 18, 1988.

35. According to the influential Nihon Keizai newspaper's front page story of December 13, 1995, 61.5 percent of salary increases for 515 firms are related to ability and 38.5 percent based on seniority. *Nihon Keizai Shimbun,* "Chingin kakumei: nenkoshugi to no ketsubetsu [Wage Revolution: The collapse of seniority-based compensation]," December 13, 1995, 1.

36. Interview with Mr. Yoshinaga, Yanagicho, August 18, 1988.

37. M. Tamada, H. Akiyama, and M. Kuwabara, *IC Kaado Shikumi to Hirogaru Sekai,* Tokyo: Denki Shoin, 1989, 27–29.

Chapter 6

1. "Asian Promise—Japanese Manufacturing," *The Economist,* June 12–19, 1993, 98–99. By 1993, Asia received 19 percent of Japanese foreign direct investment (FDI), up from 12 percent in 1990. By contrast, Japanese FDI had fallen from 46 to 40.5 percent for America and from 25 to 21 percent for Europe during the same period.

2. Martin Fransman, *Japan's Computer and Communications Industry,* Oxford: Oxford University Press, 1995, 335.

3. International Business Affairs Division, MITI, "Charts and Tables Related to Japanese Direct Investment Abroad," mimeo, April 1991.

4. Toyota Motors, for example, announced an expansion of vehicle research and development facilities to existing sales and production facilities in North America. Toyota will spend more than $220 million on this effort. Recently, Toyota Technical Center opened in California with hot and cold chambers for testing climate control systems and an anechoic chamber to measure noise and vibration. A vehicle proving facility is being built northwest of Phoenix with a planned opening in 1993. Calty Design Research, Toyota's design arm in California, is receiving new investment to upgrade its electronic drawing and color analyzing capabilities. A new parts and materials evaluation laboratory was opened in Ann Arbor, Michigan, during June 1991. *Financial Times,* June 1/2, 1991, 2. On the last point, also see Andrew Pollack, "Japan's Market Shrinking for Consumer Electronics," *New York Times,* Monday, December 27, 1993, C1–C2.

5. This may be illustrated in the following manner: not only are products and production processes technically complex but also the technical and organizational requirements for operating the manufacturing system are high. The manufacture of a photocopier machine, for example, is technically demanding, with some 900 to 1,000 parts of different purpose, precision, and material. At the same time, electrostatic, electromagnetic, opticalelectric, fiber optic, semiconductor, fine chemical (toner), precision machining, electrical mechanical, mechanical, paper handling, and other complex technologies have to be incorporated *as a system* in a high-performing photocopier.

The combination of these features may be shown as follows; obviously, it is the combination that makes the definition, regulation, and management of mental work extremely difficult. This way of illustrating the issue is based on Michael Porter's early work on globalization.

Production System

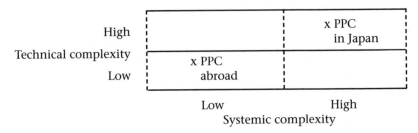

High ·············· x PPC in Japan

Technical complexity

Low ············· x PPC abroad

Low   High
Systemic complexity

6. The first comprehensive treatment of the organizational difficulties inherent in the transfer of a Japanese system of production to an overseas, offshore site is Mary Yoko Brannen's *Your Next Boss is Japanese*, New York: Oxford University Press, forthcoming.

7. These comments and others translated in this chapter were obtained in an interview with Mr. Komai on August 8, 1988.

8. For a more detailed discussion of what happened at Mitchell, South Dakota, see "Toshiba and Site-Specific Organizational Learning in the International Transfer of Technology," in Jeffrey E. Liker, W. Mark Fruin, and Paul E. Adler, eds., *Remade in America*, New York, Oxford University Press, forth-coming.

9. Kazuo Koike, *Shokuba no Rodo Kumiai to Sanka* [Workshop Unionization and Participation], Tokyo: Toyo Keizai, 1979, 222–40.

10. For a discussion of corporate-level management and engineering career paths, see Kiyonori Sakakibara and D. Eleanor Westney, "A Comparative Study of the Training, Careers, and Organization of Engineers in the Computer Industry in the United States and Japan," *Hitotsubashi Journal of Commerce and Management* 20(1) (December 1985), 1–20.

11. All of the figures given in this paper were obtained from the Purchasing Department of the Yanagicho Works in the course of fieldwork during the summers of 1987, 1988, and 1989, and the academic year 1990–91.

12. Tony Jackson, "Japanese 'Responding' Over Copiers," *Financial Times*, September 29, 1988, 6.

13. Information obtained in an interview with the deputy head of the Labor Relations Department at the Toshiba Corporation on August 15, 1988.

14. Interviews conducted on February 20 & 21, 1990. The conversations reported here are my translations of discussions conducted in Japanese. I have a good knowledge of colloquial Japanese and I have been studying Toshiba and Yanagicho for approximately six years.

15. Product integrity is a concept developed by Kim Clark and Takahiro Fujimoto in *Product Development Performance*, Boston: Harvard Business School Press, 1991.

16. Id., note #14 above.

17. The diagram below is typical of the categorization of product strategies into those that require higher and lower degrees of corporate integration and product/market integration.

Production System

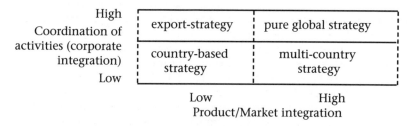

	Low	High
	export-strategy	pure global strategy
	country-based strategy	multi-country strategy

High
Coordination of
activities (corporate
integration)
Low

Low   High
Product/Market integration

18. A paper of mine that treats industry differences in product development is "Good Fences Make Good Neighbors: In Product Development as In Most Everything Else, In Japan as Most Everywhere Else," paper delivered at the workshop on Competitive Product Development, Euro-Asia Centre, IN-SEAD, June 27–29, 1991, Fontainebleau, France. The transferability problem may be more acute in some industries than in others. For example, the electronics industry, as opposed to the motor vehicle industry, is characterized by:

- a greater number and variety of products,
- shorter product life-cycles,
- more technical discontinuity between product generations,
- more intrafirm rivalry between different design-build-test teams working in similar market segments,
- more fungibility in core and enabling technologies,
- more markets (intermediate and final) for more products.

19. The best known treatment of Japanese transplant success is J. Womack, D. Jones, and D. Roos, *The Machine That Changed the World*, New York: Rawson Associates (Macmillan), 1991. Also for consumer electronics, see Malcolm Trevor, *Toshiba's New British Company*, London: Policy Studies Institute, 1988.

20. Arguably, transfer and implantation are easier for consumer electronics, such as televisions, compared with PPCs. Since 70 percent of the average TV cost is in raw materials, localization efforts are simplified to the extent that acceptable materials are available. In the case of PPCs, however, less than 20 percent of the average cost is in materials. Most of it is in cogs, wheels, drums, pulleys, spindles, and tubes of high tensile metals and space-age plastics. Strengthening relationships with local suppliers in the former case is entirely different than in the latter. Likewise, getting closer to the market in one case is rather different than in the other. See, "Putting Collaboration into Sharper Focus," *Financial Times*, May 17, 1991, 16.

21. Lawrence E. Fouraker and John M. Stopford, "Organizational Structure and Multinational Strategy," *Administrative Science Quarterly* 13(1) (June 1968), 47–64.

22. "Overseas Investment by Japanese Companies Hits Another Record High," *Business Asia*, June 20, 1988, 200–201.

Chapter 7

1. "Low" in all these instances means not only lower costs but lower productivity and output as well. For companies like Toshiba, however, focused on product-making as the principal means of profit-making, higher costs must be offset with the manufacture of higher-value-added products. In industries like microelectronics, telecommunications, robotics, and automation systems, products are inherently complex in terms of their hardware and software designs, performance features, number of embedded technologies, and processing requirements. These industries demand production and sales engineering systems of great complexity, precision, and reliability.

In these industries, fast-follower strategies, or a tactic of imitating rather than innovating products at the leading edge of practice, is no longer viable. Even though fast-follower strategies generally raised Japanese companies' performance to international standards during the 1960s, 70s, and 80s, today markets change too fast, products and processes are too complicated, systems engineering requirements too daunting.

2. William E. Fruhan, Jr., "The NPV (Net Present Value) Model of Strategy—The Shareholder Value Model," *Financial Strategy: Studies in the Creation, Transfer, and Destruction of Shareholder Value*, Homewood, Ill.: Richard D. Irwin, 1979.

3. Kim Clark and Takahiro Fujimoto, *Product Development Performance*, Boston: Harvard Business School, 1991.

4. Interview with Mr. Tetsuya Yamamoto, Senior Manager, Logistics, Toshiba Corporation, Shibaura, Tokyo, April 3, 1992.

5. G. Dan Hutcheson and Jerry D. Hutcheson, "Technology and Economics in the Semiconductor Industry," *Scientific American* 274(1) (January 1996), 54–59.

6. W. Mark Fruin, *The Japanese Enterprise System*, Oxford: Clarendon Press, 1992.

7. For a long discussion of the evolution of focal factories in Japan, see *The Japanese Enterprise System*.

8. Of course, the organizational interdependence of factory, firm, and interfirm network was the theme of *The Japanese Enterprise System*.

9. Please note, however, that a 10 percent rate of revenue growth was considered "lackluster" in Japan during the 1970s and into the 1980s. This rate of growth would make many a large Western firm smile, especially during the oil-shocked 70s. The information contained in this section comes from Nomura Research Institute, TOSHIBA, *NRI Report*, Feb. 5, 1990, 52 pp.

For Toshiba, as well as for every other large firm in Japan, sales and revenues need to be evaluated on both a consolidated and unconsolidated basis. For the 1989 fiscal year, for example, Toshiba's sales on a consolidated basis (after eliminating double counting of internal transactions) were 1.3 times greater than for Toshiba proper, while profits were 1.95 times higher. However, these figures are for one year only, 1989, and given the nature of the Japanese enterprise system where interorganizational ties bind factories, firms, and interfirm affiliates, there is likely to be a regular oscillation in rates of unconsolidated and consolidated sales, revenues, and profits, depending on the degree to which interorganizational resources have been mobilized in preceding years. Also, the policies of core companies, like To-shiba, toward affiliates are not the same from year to year and these pow-erfully affect accounting results.

10. Interview with Seiichi Takayanagi, senior advisor, Toshiba Corpora-tion, Hammatsucho, Tokyo, September 20, 1994.

11. Nomura Research Institute, *NRI Report.*

12. Michael Cusumano, *Japan's Software Factories*, New York: Oxford University Press, 1991.

13. Mary Douglas, *How Institutions Think*, Syracuse: Syracuse University Press, 1986.

14. *The Japanese Enterprise System* establishes the argument for seeing the interrelations between factory, firm, and interfirm network as an orga-nizational system.

15. M. Mitchell Waldrop, *Complexity*, New York: Simon & Schuster, 1992.

16. Martin Kenney and Richard Florida, *Beyond Mass Production*, New York: Oxford University Press, 1993. Kenney and Florida posit shop-floor-mediated, incremental innovation as the strength of Japan's manufactur-ing. While shop floor innovation is important, I am reluctant to see it as an independent variable. It is dependent on managerial, organizational, and to a lesser degree, technical factors, in my opinion.

17. Robert H. Hayes, Steven C. Wheelwright, and Kim B. Clark, *Dynamic Manufacturing—Creating the Learning Organization*, New York: The Free Press, 1988.

18. Rather than discuss each of the prevailing models of factory pro-duction separately, the following list may be helpful in suggesting *what Knowledge Works are not:*

- not simply centers of technical diffusion, even though this is what happens within factories as well as within networks of suppliers and affiliates cooperating with Knowledge Works;
- not plants within plants (PWP), due to the integration of resources within-factories, and the sharing of learning within the network of suppliers and affiliates;
- not simply flexible factories where a lot of different products are made and various assembly and manufacturing process changes occur;
- not centers of excellence (à la Ford Motors);
- not just a result of plant and district agglomeration effects, although these happen in predictable ways;
- not only concerned with learning economies, even if these are con-sciously planned for and important in Knowledge Works.

19. Nihon Keizai Shimbun, "Price is the Key to Toyota's New Luxury Car for the US Market," *Nihon Keizai Shimbun*, September 14, 1994, 13.

20. Dr. Seiichi Takayanagi, senior adviser, Toshiba Corporation, public presentation, Almasa, Sweden, August 19, 1994. According to Dr. Takayan-agi, former director of R & D Management at Toshiba, Toshiba engineers and managers would like to realize the organizational focus that Toyota does. But Toshiba's system of manufacturing organization with 90 percent of university trained R & D workers in Knowledge Works makes this diffi-cult. It does, however, allow Toshiba to develop many more domains of

technical capability across many more product and market segments than Toyota does.

21. See Banri Asanuma for a description of Mitsubishi Electric's factory system in "Setsubi Toshi Kettei no Prosesu to Kijun (2) [Decision-Making Processes and Standards for Plant and Equipment Investments]," *Keizai Ronso* [Economic Papers], Kyoto University, Department of Economics, 130 (5/6) (November–December 1982).

Fungibility refers to the ease with which technical resources may be employed for alternative uses. Fungibility is analogous to flexibility but different; it refers to something organic and very basic in organizational development, while flexibility is closer to something mechanistic, more physical than biological. The point is that certain technologies are more fungible than others. Petroleum cracking technology, or the chemical decomposition of petroleum, yields any number of intermediate and end products, such as kerosene, gasoline, diesel, and aviation fuel. Compare this with steel making technology where intermediate and end products are fewer. Thus, chemical engineering technologies are more fungible than metal refining technologies; electronic technologies more fungible than mechanical technologies. Electronic technologies are among the most fungible of all technologies, allowing Toshiba's Knowledge Works a breadth and depth of endeavor that firms in other industries with less fungible technologies can only admire.

22. W. Mark Fruin and Toshihiro Nishiguchi, "Supplying the Toyota Production System," in Bruce Kogut, ed., *Country Competitiveness*, New York: Oxford University Press, 1994, 225–46. Also, see Introduction and the appendices of *The Japanese Enterprise System*, Ibid.

23. The relationship between design and function in microbiology is shown in Matt Ridley's *The Red Queen*, New York: Penguin Books, 1995. The parallels with organization theory are fascinating. For a discussion of the relationship between institutional imitation and innovation, see Eleanor Westney, *Imitation and Innovation*, Cambridge: Harvard University Press, 1987, and Nathan Rosenberg and Edward Steinmueller, "Why Are Americans Such Poor Imitators?" *American Economic Review* 78(2) (May 1988), 229–34.

24. Economic Planning Agency, *Keizai Hakusho*, Tokyo: Finance Ministry Press, 1990, 143–255, and the National Science Foundation, *The Science and Technology Resources of Japan: A Comparison with the United States*, NSF 88-318, Wash., D.C., 1988.

25. Michael L. Dertouzos, Richard K. Lester, and Robert M. Solow, *Made in America*, New York: Harper Perennial Edition, 1989.

26. Factories were recognized as complex, social economic systems long ago. I am simply picking up on themes that have been neglected during the last three or four decades. A very important statement of the problem appeared in 1947 as Vol. 4 of the Yankee City Series: W. Lad Wagner and J.O. Low, *The Social System of the Modern Factory*, New Haven: Yale University Press, 1947. For a Swedish, prolabor view of factories as complex social economic systems, see Stephen Aguren and Jan Edgren, *New Factories—Job Design through Factory Planning in Sweden*, Stockholm: Swedish Employers Confederation, 1980.

27. Ray Rees, "The Share Economy and Alternative Systems of Workers' Remuneration: A Review of Weitzman and Meade," *Journal of the Japanese and International Economies* 4(2) (June 1980), 194–214.

28. Douglass C. North, *Institutions*, 28, 34. Also, Whittaker compares Japanese and English small factories during processes of technology upgrading and workplace automation. See D. H. Whittaker, *Managing Innovation*, Cambridge: Cambridge University Press, 1990.

Bibliography

Works in English

Abegglen, James. *The Japanese Factory*. Homewood, Ill.: Free Press, 1958.

Abegglen, James, and George Stalk. *Kaisha-The Japanese Corporation*. New York: Basic Books, 1985.

Abo, Tetsuo. *The Hybrid Factory*. New York: Oxford University Press, 1992.

Adler, Paul. "The Learning Bureaucracy: New United Motor Manufacturing, Inc." In Barry M. Staw and Larry L. Cummings, eds., *Research in Organizational Behavior*. Greenwich, Conn.: JAI Press, 1993.

Adler, Paul, and Robert Cole. "Designed for Learning: A Tale of Two Auto Plants." *Sloan Management Review* (Spring 1993): 85–94.

Aguren, Stephen, and Jan Edgren. *New Factories-Job Design through Factory Planning in Sweden*. Stockholm: Swedish Employers' Confederation, 1980.

Aoki, Masahiko. *Information, Incentives, and Bargaining in the Japanese Economy*. Cambridge: Cambridge University Press, 1988.

———. "Horizontal vs. Vertical Information Structure of the Firm." *The American Economic Review* 76–5 (Dec. 1986): 971–83.

Aoki, Masahiko, and Ronald Dore. *The Japanese Firm-Sources of Competitive Strength*. Oxford: Clarendon Press, 1994.

Argyris, C. "Teaching smart people how to learn." *Harvard Business Review* 69–3 (1991): 99–109.

Arrow, Kenneth. *Limits of Organization*. New York: W. W. Norton, 1974.

Asanuma, Banri. "Transactional Structure of Parts Supply in the Japanese Automobile and Electric Machinery Industries: A Comparative Analysis." *Technical Report No. 1*, Socio-Economic Systems Research Project. Kyoto: Kyoto University, July 1985.

———. "Manufacturer-Supplier Relationships in Japan and the Concept of Relation-Specific Skill." *Journal of the Japanese and International Economics* 3 (1989): 1–30.

Axelrod, Robert. *The Evolution of Cooperation*. New York: Basic Books, 1984.

Azumi, Koya, and Frank Hull. "Inventive Payoff from R & D in Japanese Industry: Convergence with the West?" *IEEE Transactions on Engineering Management* 37–1 (February 1990): 3–9.

Bellah, Robert N., Richard Madsen, William M. Sullivan, Ann Swindler, and Steven M. Tipton. *Habits of the Heart*. Berkeley: University of California Press, 1985.

Boas, Franz. *Race, Language, and Culture*, Chicago: University of Chicago Press, 1982.

Bradbury, Ray. *Fahrenheit 451*. New York: Ballantine Books, 1953.

Bradsher, Keith. "Skilled Workers Watch Their Jobs Migrate Overseas." *New York Times*, August 28, 1995, A1 and D6.

Brannen, Mary Yoko. *Your Next Boss is Japanese*. University of Massachusetts, Amherst. Department of Anthropology and School of Management joint Dissertation, 1994.

——. *Negotiated Culture and Recontextualization*. New York: Oxford University Press, forthcoming.

Burgelman, Robert A., and Modesto A. Maidique. *Strategic Management of Technology and Innovation*. Homewood, Ill.: Irwin, 1988.

Business Asia. "Overseas Investment by Japanese Companies Hits Another Record High." June 20, 1988, 200–201.

Chandler, Alfred D., Jr. *Strategy and Structure: Chapters in the History of Industrial Enterprise*. Cambridge: MIT Press, 1962.

——. Jr. *Scale and Scope*, Cambridge: Harvard University Press, 1990.

Chen, Bill. Mimeo to Kouji Kuramochi, March 22, 1989; Visa International's San Mateo, California, offices.

Choe, Wolhee Choe. *Towards an Aesthetic Criticism of Technology*. New York and Bern: Peter Lang, 1989.

Clark, Kim B., Robert H. Hayes, and Christopher Lorenz, eds. *The Uneasy Alliance: Managing the Productivity-Technology Dilemma*. Boston: Harvard Business School Press, 1985.

Clark, Kim B. and Takahiro Fujimoto. *Product Development Performance*. Boston: Harvard Business School Press, 1991.

Cole, Robert E. *Japanese Blue Collar*. Berkeley: University of California Press, 1971.

——. *Work, Mobility and Participation*. Berkeley: University of California Press, 1979.

——. *Strategies for Learning*. Berkeley: University of California Press, 1989.

Crozier, Michel. *The Bureaucratic Phenomenon*. London: Tavistock Institute, 1964.

Cusumano, Michael A. *Japan's Software Factories: A Challenge to U.S. Management*. New York: Oxford University Press, 1991.

D'Aveni, Richard A., with Robert Gunther. *Hyper-competitive Rivalries*. New York: The Free Press, 1994.

Dertouzos, Michael L., Richard K. Lester, and Robert M. Solow. *Made in America*. New York: Harper Perennial Edition, 1989.

Dick, Philip K. *Do Androids Dream of Electric Sheep*. London: Panther, 1972.

——. *Dr. Bloodmoney: or, How We Got Along After the Bomb*, Boston: Gregg Press, 1980.

DiMaggio, Paul J., and Walter W. Powell. "The Iron Cage Revisited: Institutional Isomorphism and Collective Rationality in Organizational Fields." *American Sociological Review* 48 (April 1983): 147–60.

Dore, Ronald. *British Factory-Japanese Factory*. Berkeley: University of California Press, 1973.

——. *Flexible Rigidities*. Stanford: Stanford University Press, 1986.

Douglas, Mary. *How Institutions Think*. Syracuse: Syracuse University Press, 1986.

Douglas, Mary, and Aaron Wildavsky. *Risk and Culture: An Essay on the se-*

lection of Technological and Environmental Dangers. Berkeley: University of California Press.

Drucker, Peter. *Post-Capitalist Society*. New York: Harper Business, 1993.

Dyer, G. W., Jr., and A. L. Wilkins. "Better stories, not better constructs, to generate better theory: A rejoinder to Eisenhardt, *Academy of Management Review*, 16-3 (1991): 620–27.

The Economist. "Asian Promise-Japanese manufacturing." June 12–19, 1993, 98–99.

Eisenhardt, K. M. "Better stories and better constructs: The case for rigor and comparative logic." *Academy of Management Review*, 16–3 (1991): 613–19.

Eisenhardt, Kathleen M., and Benham M. Tabrizi. "Accelerating Adaptive Processes: Product Innovation in the Global Computer Industry." *Administrative Science Quarterly* 40–1 (March 1995): 84–110.

Etzioni, Amitai. *Political Unification: A Comparative Study of Leaders and Forces*. New York: Holt, Rinehart, and Winston, 1965.

———. *A Comparative Analysis of Complex Organizations*. Rev. ed. New York: Free Press, 1975.

Ferdows, Kasra, ed. *Managing International Manufacturing*. Amsterdam: North Holland, 1989.

Financial Times. "Putting collaboration into sharper focus." May 17, 1991, 16.

Finney, Paul Burnham. "Executives on the go are the main buyers of portable computers." *Business Travel*, New York Times, January 17, 1996, C3.

Fouraker, Lawrence E., and John M. Stopford. "Organizational Structure and Multinational Strategy." *Administrative Science Quarterly* 13–1 (June 1968).

Franko, Lawrence G., "The Japanese Juggernaut Rolls On," *Sloan Management Review*, 37–2 (Winter 1996), 103–09.

Fransman, Martin. *The Market and Beyond*. Cambridge: Cambridge University Press, 1990.

———. *Japan's Communications and Computer Industry*. Oxford: Oxford University Press, 1995.

Freeman, Christopher. *Technology Policy and Economic Performance*. London: Pinters, 1987.

Fruhan, William E., Jr. "The NPV (Net Present Value) Model of Strategy-The Shareholder Value Model." In *Financial Strategy: Studies in the Creation, Transfer, and Destruction of Shareholder Value*. Homewood, Ill.: Richard D. Irwin, 1979.

Fruin, W. Mark. *The Japanese Enteprise System-Competitive Strategies and Co-operative Structures*. Oxford: Clarendon Press, 1992.

———. "Competing the Old-Fashioned Way: Localizing and Integrating Knowledge Resources in Fast-to-Market Competition." In J. Liker, J. Ettlie, and J. Campbell, eds. *Engineered in Japan: Technology Management Practices in Japan*. New York: Oxford University Press, 1995.

———. "Good Fences Make Good Neighbors-Organizational Property Rights and Permeabiity in Product Development Strategies in Japan." Paper delivered at the Workshop on Competitive Product Development, Euro-Asia Centre, INSEAD, June 27–29, 1991, Fontainebleau, France.

Fruin, W. Mark, and Toshihiro Nishiguchi. "Supplying the Toyota Production System." In Bruce Kogut, ed., *Country Competitiveness*. New York: Oxford University Press, 1992.

Fucini, J., and S. Fucini. *Working for the Japanese: Inside Mazda's American Auto Plant*. New York: The Free Press, 1990;

Galbraith, Jay R. *Designing Complex Organizations*. Reading, Mass.: Addison-Wesley, 1973.

Gerlach, Michael. *Alliance Capitalism*. Berkeley: University of California Press, 1992.

Gillespie, Richard. *Manufacturing Knowledge*. New York: Cambridge University Press, 1991.

Graham, Margaret, and Betty Pruett. *R & D for Industry*. Cambridge: Cambridge University Press, 1991.

Gresser, Julian. *Piloting Through Chaos*. Sausalito: Five Rings Press, 1995.

Griliches, Zvi. "R & D and the Productivity Slowdown." *American Economic Review* 70–2 (1980): 343–48.

Hamel, Gary, and C. K. Prahalad. *Competing for the Future*. Boston: Harvard Business School Press, 1994.

Harmon, Roy L., and Leroy D. Peterson. *Reinventing The Factory*. New York: Basic Books, 1990.

Hayes, Robert H., Steven C. Wheelwright, and Kim B. Clark. *Dynamic Manufacturing-Creating the Learning Organization*. New York: The Free Press, 1988.

Hayes, Robert H. and Kim B. Clark. "Why Some Factories are More Productive than Others." *Harvard Business Review* (September/October 1986): 66–73.

Hayes, Robert H., and Ranchandran Jaikumar. "Manufacturing's Crisis: New Technologies, Obsolete Organizations." *Harvard Business Review* (September–October 1988): 77–85.

Helper, S. "An Exit-Voice Analysis of Supplier Relations." In R. Coughlin, ed., *Morality, Rationality and Efficiency: New Perspectives on Socio-Economics*. New York: M. E. Sharpe, 1991, 355–72.

Hill, Charles W. L. "Cooperation, Opportunism, and The Invisible Hand: Implications for Transaction Cost Theory." *Academy of Management Review* 15–3 (1992): 500–13.

Hippel, E. von. *Sources of Innovation*. Oxford: Oxford University Press, 1988.

Hoskisson, Robert E., and Michael A. Hitt. "Strategic Control Systems and Relative R & D Investment in Large Multiproduct Firms." *Strategic Management Journal* 9 (1988): 605–21.

Hull, Frank M., Jerald Hage, and Koya Azumi. "R & D Management Strategies: America versus Japan." IEEE Transactions on Engineering Management 32–2 (May 1985): 78–83.

Hull, Frank M., and Koya Azumi. "Teamwork in Japanese and U.S. Labs." *Research Technology Management* 32–6 (Nov–Dec 1989): 21–26.

Hutcheson, G. Dan, and Jerry D. Hutcheson. "Technology and Economics in the Semiconductor Industry." *Scientific American* 274–1 (January 1996): 54–59.

Imai, Ken'ichi, Ikujiro Nonaka, and Hirotake Takeuchi. "Managing the New Product Development Process: How Japanese Companies Learn and Unlearn." In Kim B. Clark, Robert H. Hayes, and Christopher Lorenz, eds., *The Uneasy Alliance*. Boston: Harvard Business School, 1985.

Jackson, Kenneth T. *The Encyclopedia of New York City*. New Haven and New York: Yale University Press and the New York Historical Society, 1995.

Jackson, Tony. "Japanese 'responding' over copiers." *Financial Times*, September 29, 1988, 6.

Jacobs, Jane. *Cities and the Wealth of Nations: Principles of Economic Life*. New York: Random House, 1984.

Juran, Joseph. "Why Quality in Japan." *Washington Post*, op/ed commentary page, Aug. 15, 1993.

Itami, Hiroyuki, and Thomas W. Roehl. *Mobilizing Invisible Assets*. Cambridge: Harvard University Press, 1987.

Kanter, Rosabeth Moss. *The Change Masters*. New York: Simon & Schuster, 1983.

Kawasaki, Seiichi, and John McMillan. "The Design of Contracts: Evidence from Japanese Subcontracting." *The Journal of Japanese and International Economies* 1–3 (1987): 327–49.

Kenney, Martin, and Richard Florida. *Beyond Mass Production*, New York: Oxford University Press, 1992.

———. "The Organization and Geography of Japanese R & D: Results from a survey of Japanese electronics and biotechnology firms." *Research Policy* 23 (1994): 305–23.

Kidder, J. T. *The Soul of a New Machine*. New York: Avon Books, 1981.

Kodama, Fumio. *Analyzing Japanese High Technologies: The Techno-Paradigm Shift*. London: Pinter Publishers, 1991.

Kuznets, Simon. *Modern Economic Growth: Rate, Structure, Spread*. New Haven: Yale University Press, 1966.

Langlois, Richard N. "Transaction Cost in Real Time." *Industrial and Corporate Change* 1–1 (1992): 99–127.

Leibenstein, Harvey. "Aspects of the X-efficiency Theory of the Firm." *Bell Journal of Economics* 6–2 (Fall 1975): 580–606.

———. *Beyond Economic Man*. Cambridge: Harvard University Press, 1976.

Levitt, B., and J. G. March. "Organizational Learning." *Annual Review of Sociology* 14 (1988): 319–40.

Liker, Jeffrey K., John E. Ettlie, and John C. Campbell. *Engineered in Japan: Japan Technology Management Practices*. New York: Oxford University Press, 1995.

Lillrank, Paul, and Noriaki Kano. *Continuous Improvement: Quality Control Circles in Japanese Industry*. Ann Arbor: Center for Japanese Studies, University of Michigan, 1989.

Lincoln, James R., and Arne L. Kalleberg. *Culture, Control, and Commitment*. Cambridge: Cambridge University Press, 1990.

Lincoln, James R., Michael L. Gerlach, and Christina Ahmadjian. "Keiretsu Networks and Corporate Performance in Japan." forthcoming in *American Sociological Review*.

Link, Albert N. "Basic Research and Productivity Increase in Manufacturing: Additional Evidence." *American Economic Review* 71– 5 (1981): 1,111–12.

Los Angeles Times. "United Technologies to Cut 13,900 Jobs in Restructuring." January 22, 1992, D1–2.

———. "Unrest: Urban Turmoil Throughout the World." May 25, 1992, A1, 36–7.

Maccoby, Michael. "Forces for Increased Organizational Competition: Implications for Leadership and Organizational Competence." In Daniel B. Fishman and Cary Cherniss, eds., *The Human Side of Corporate Competitiveness*. Newbury Park and London: Sage Publications, 1990.

Mandell, Lewis. *The Credit Card Industry*. Boston: Twayne Publishers, 1990.

Mansfield, Edwin. "Basic Research and Productivity Increase in Manufacturing." *American Economic Review* 70–5 (1980): 863–73.

———. "The Speed and Cost of Industrial Innovation in Japan and the United States: External vs. Internal Technology." *Management Science* 34–10 (Oct. 1988): 1,157–68.

———. "Competitiveness of American Manufacturing: Technological and Productivity Factors." *Managerial and Decision Economics*, Special Issue (1989): 1–2.

———. "Technological Change in Robotics: Japan and the United States," *Managerial and Decision Economics*, Special Issue (1989): 19–26.

Marx, Karl, and Frederick Engels. *Manifesto of the Communist Party*. English trans. Moscow: Foreign Languages Publishing House, 1955.

Nadiri, M. Ishaq. "Sectoral Productivity Slowdown." *American Economic Review* 70–2 (1980): 349–52.

———. "Contributions and Determinants of Research and Development Expenditures in the U.S. Manufacturing Industries." In G. M. von Furstenberg, ed., *Capital, Efficiency and Growth*. Cambridge: Ballinger 1980.

National Science Foundation. *The Science and Technology Resources of Japan: A Comparison with the United States*. NSF 88–318. Washington, D. C., 1988.

Nelson, Richard R., and Sidney G. Winter. *An Evolutionary Theory of Economic Change*. Cambridge: Belknap Press, Harvard University Press, 1982.

Nielsen, Richard P. "Cooperative Strategy." *Strategic Management Journal* 9 (1988): 475–92.

Nishiguchi, Toshihiro. *Strategic Industrial Sourcing*. New York: Oxford University Press, 1994.

Nohria, Nitin, and Robert G. Eccles. *Networks and Organizations*. Boston: Harvard Business School Press, 1992.

Nonaka, Ikujiro, and Hirotaka Takeuchi. *The Knowledge-Creating Company*. New York: Oxford University Press, 1995.

North, Douglass C. *Institutions, Institutional Change, and Economic Performance*. Cambridge: Cambridge University Press, 1990.

Odagiri, Hiroyuki. "Research Activity, Output Growth, and Productivity Increase in Japanese Manufacturing Industries." *Research Policy* 14 (1985): 117–30.

Odagiri, Hiroyuki, and Hitoshi Iwata. "The Impact of R&D on Productivity Increase in Japanese Manufacturing Companies." *Research Policy* 15 (1986): 13–19.

Ohmae, Ken'ichi. *The Borderless World*. New York: Harper Business, 1990.

Oliver, C. "Determinants on interorganizational relationships: integration and future directions." *Academy of Management Review* 15–2: 241–65.

Oniki, Hajime. "On the Cost of Deintegrating Information Networks." In C.

Antonelli, ed., *The Economics of Information Networks*. Amsterdam: Elsevier Science Publishers, 1992.

Penrose, Edith. *The Theory of the Growth of the Firm*. White Plains, N.Y.: M. E. Sharpe, 1980.

Piore, Michael, and Charles Sabel. *The Second Industrial Divide*. New York: Basic Books, 1984.

Pollack, Andrew. "Japan Eases 57 Varieties Marketing." *The New York Times*. October 15, 1992, D1, D7.

———. "Japan's Market Shrinking for Consumer Electronics." *The New York Times*. December 27, 1993, C1–C2.

———. "Stunning Changes in Japan's Economy." *The New York Times*. October 23, 1994, section 3, 1.

Powell, Walter W. "Neither Market nor Hierarchy: Network Forms of Organization." *Research in Organizational Behavior* 12 (1990): 295–336.

Prahalad, C. K., and Gary Hamel. "The Core Competence of the Corporation." *Harvard Business Review* (May–June 1990): 79–91.

Quinn, James Brian, Henry Mintzberg, and Robert M. James, eds. *The Strategy Process*. New York: Prentice Hall, 1988.

Rees, Ray. "The Share Economy and Alternative Systems of Workers' Remuneration: A Review of Weitzman and Meade." *Journal of the Japanese and International Economies* 4–2 (June 1980): 194–214.

Reich, Robert. *The Work of Nations*. New York: Vintage Books, 1992.

Report to Visa International. Memo, September 1, 1988 meeting. San Mateo, California.

Ridley, Matt. *The Red Queen*. New York: Penguin Books, 1995.

Rosenberg, Nathan, and W. Edward Steinmueller. "Why Are Americans Such Poor Imitators?" *American Economic Review* 78–2 (May 1988): 229–34.

Sabbagh, Karl. *21st Century Jet*. New York: Scribners, 1966.

Sakakibara, Kiyonori, and D. Eleanor Westney. "A Comparative Study of the Training, Careers, and Organization of Engineers in the Computer Industry in the United States and Japan." *Hitotsubashi Journal of Commerce and Management* 20–1 (December 1985): 1–0.

Sato, Kazuo. "Saving and Investment." In Kozo Yamamura and Yasukichi Yasuba, eds., *The Political Economy of Japan*. Vol. 1. The Domestic Transformation. Stanford: Stanford Univ Press, 1987.

Schein, Edgar H. *Organizational Culture and Leadership*. San Francisco: Jossey Bass, 1985.

———. "Organizational Culture." MIT Working Paper #2088–88, December 1988.

Schmenner, Roger W. "Behind Labor Productivity Gains in the Factory." *Journal of Manufacturing and Operations Management* 1–4 (Winter 1988): 323–38.

Schmenner, Roger W., and Boo Ho Rho. "An International Comparison of Factory Productivity." *International Journal of Operations and Production Management* 10–4 (1990): 16–31.

Scott, W. R. *Organizations: Rational, Natural, and Open Systems*. 2nd Ed. Englewood Cliffs, N.J.: Prentice Hall, 1987.

Shane, S., S. Venkataraman, and I. MacMillan. "The Effects of Cultural Dif-

ferences on New Technology Championing Behavior within Firms."
Journal of High Technology Management Research 5–2 (Fall 1994): 163–82.

Sheard, Paul, ed. *International Adjustment and the Japanese Firm*. St. Leonards,
NSW: Allen & Unwin, 1992.

———. "Japanese Corporate Boards and the Role of Bank Directors." In W.
Mark Fruin, ed., *Networks and Markets: Pacific Rim Strategies* forthcoming
1997.

Shirai, Taishiro, ed. *Contemporary Industrial Relations in Japan*. Madison: University of Wisconsin Press, 1983.

Skinner, Wickham. "The Focused Factory." *Harvard Business Review* (May–
June 1974): 113–21.

Smith, Thomas C. *Native Sources of Japanese Industrialization, 1750–1920*.
Berkeley: University of California Press, 1988.

Smitka, Michael J. *Competitive Ties*. New York: Columbia University Press,
1991.

Spence, A. Michael. "The Learning Curve and Competition." *Bell Journal of
Economics* 12–1 (1981): 49–70.

Stalk, George, and Thomas M. Hout. *Competing Against Time*. New York: The
Free Press, 1990.

Stewart, James B. *Den of Thieves*. New York: Simon & Schuster, 1991.

Teece, David J. "Towards an Economic Theory of the Multiproduct Firm."
Journal of Economic Behavior and Organization 3 (March 1982): 39–64.

———. "Perspectives on Alfred Chandler's Scale and Scope." *Journal of Economic Literature* 31–1 (March 1993): 199–225.

Toshiba Newsletter. "Toshiba To Produce Medical Electronics & Telecommunications Equipment in U.S." No. 281. Tokyo, Dec. 1985.

Trevor, Malcolm. *Toshiba's New British Company*. London: Policy Studies Institute, 1988.

Turner, Christena. *Breaking the Silence*. Stanford University, Department of
Anthropology Ph.D. Dissertation, 1988.

Tushman, Michael L., and William L. Moore, eds. *Readings in the Management of Innovation*. New York: Harper Business, 1988.

Uchimaru, Kiyoshi, Susumu Okamoto, and Buntera Kurahara. *TQM for Technical Groups*. Portland, Ore.: Productivity Press, 1993.

Utterback, J. M., and W. J. Abernathy. "A Dynamic Model of Process and
Product Innovation." *Omega* 3–6 (1975): 639–56.

Van de Ven, Andrew H., Harold L. Angle, and Marshall Scott Poole. *Research
on the Management of Innovation*. New York: Harper & Row, 1989.

Van de Ven, Andrew H., and G. Walker. "The dynamics of interorganizational coordination." *Administrative Science Quarterly* 29–4: 598–621.

Velocci, Anthony L. "Boeing, Labor Talks Ripple Industrywide." *Aviation
Week & Space Technology*, August 14, 1995, 20–21.

Verne, Jules. *The Annotated Guide to Jules Verne: Twenty Thousands Leagues
Under the Sea*. New York: Crowell, 1976.

Visa International. Internal memo. August 7, 1987.

———. Progress Report-Key Component Improvements. Item 6.1. September 24, 1987.

———. Progress Report. Internal company documents. March 1, 1988 and
June 21, 1988.

Wagner, W. Lad, and J. O. Low. *The Social System of the Modern Factory*. Vol. 4. New Haven: Yale University Press, Yankee City Series, 1947.

Waldman, David A., and Leanne E. Atwater. "The Nature of Effective Leadership and Championing Processes at Different Levels in a R & D Hierarchy." *The Journal of High Technology Management* 5–2 (Fall 1994): 233–46.

Waldrop, M. Mitchell. *Complexity*. New York: Simon & Schuster, 1992.

Wall Street Journal. "Long Road Ahead: Detroit Needs Big Overhaul to Match Japanese." January 10, 1992, A1, 6.

Wells, H. G. *The Time Machine*. New York: Heritage Press, 1964.

Westney, D. Eleanor. *Imitation and Innovation*. Cambridge: Harvard University Press, 1987.

Whittaker, D. H. *Managing Innovation-A study of British and Japanese factories*. Cambridge: Cambridge University Press, 1990.

Williamson, Oliver E. *Markets and Hierarchies: Analysis and Antitrust Implications*. New York: Free Press, 1975.

———. *Economic Institutions of Capitalism*. New York: Free Press, 1985.

Willman, Paul. *Technological Change, Collective Bargaining, and Industrial Efficiency*. Oxford: Clarendon Press, 1986.

Womack, J., D. Jones, and D. Roos. *The Machine That Changed the World*. New York: Rawson Associates (Macmillan), 1991.

Yamamura, Kozo. "Success that Soured: Administrative Guidance and Cartels in Japan." In Kozo Yamamura, ed., *Policy and Trade Issues of the Japanese Economy*. Seattle: University of Washington Press, 1982.

Works in Japanese

Akazawa, Motoaki. *Toshiba no Nijuiseiki Senryaku: Niusofutoka e no Chosen* [Toshiba's 21st-Century Strategy: The Challenge of a New Soft Transformation]. Tokyo: Nihon Noritsu Kyokai, 1991.

Asanuma, Banri. "Setsubi Toshi Kettei no Prosesu to Kijun (2) [Decision-Making Processes and Standards for Plant and Equipment Investments]," *Keizai Ronso* 130–5/6 (Nov.–Dec. 1982): 23–44.

———. "Yasashi Keizaigaku." *Nihon Keizai Shimbun*. February 21–25, 1984.

———. "The Structure of Parts Transactions in the Automotive Industry-Mechanisms of Adjustment and Innovative Adaptation." *Gendai Keizai* 59 (Summer 1984).

Daily Yomiuri. "Technology trade runs surplus in fiscal 1993." January 5, 1995, 12.

Economic Planning Agency. *Keizai Hakusho* [Economic White Paper]. Tokyo: Finance Ministry Press, 1990.

Hamaguchi, Eshun, ed. *Nihonteki shudanshugi* [Japanese-style Groupism]. Tokyo: Yuihikaku, 1982.

Hirao, Koji, Yukio Honda, and Tatsuhiro Masuda. *Kenkyu Kaihatsukei Kigyo no Shinseicho Senryaku* [New Growth Strategies for Research Intensive Companies]. Tokyo: Toyo Keizai, 1985.

Igarashi, Ryo. *Tahinshushoryo Seisan Kigyo no Keiei Kanri Kaizen* [Improving the Management of Low-Volume, High-Product Variety Firms]. Tokyo: Nikkan Kogyo, 1984.

Imai, Ken'ichi, ed. *Inobeshon to Soshiki* [Innovation and Organization]. Tokyo: Toyo Keizai, 1986.

Iwai, Masakazu. *Gijutsushatachi no Shoshudan Katsudo* [Small Group Activities for Technologists]. Tokyo: Daiyamondo, 1985.

Japan Management Association. *TP Manejimento '91 Nendo* [1991 Yearbook of TP Management]. Tokyo: JMA, 1991.

Kikuchi, Jun'ichi, Shunsuke Mori, Yasunori Baba, and Yoshiki Morino. *Kenkyu Kaihatsu no Dainamikkusu* [The Dynamics of Research and Development]. National Institute of Science and Technology Policy (NISTEP), Science and Technology Agency. Report No. 14, September 1990.

Kogure, Masao. *Nihon no TQC* [Japan's TQC]. Tokyo: Nikajiren, 1988.

Koike, Kazuo. *Shokuba no Rodo Kumiai to Sanka* [Workshop Unionization and Participation]. Tokyo: Toyo Keizai, 1979.

Kosei Torihiki Iinkai. *Kyodo Kenkyu Kaihatsu to Kyoso Seisaku* [Cooperative R & D and Competition Policy]. Tokyo: Kyosei, 1990.

Kurahara, Bunteru, Kiyoshi Uchimaru, and Susumu Okamoto. *Gijutsu Shudan no TQC* [TQC for Engineers]. Tokyo: Nikajiren, 1990. (See same publication listed in English language section under Kiyoshi Uchimaru.)

MITI, International Business Affairs Division. "Charts and Tables Related to Japanese Direct Investment Abroad." *Mimeo*, April 1991.

Miwa, Shingo. *Toshiba no Dainabukku Senryaku* [Toshiba's Dynabook Strategy]. Tokyo: Soft Bank, 1990.

Monden, Yasuhiro, with Taiichi Ohno. *Toyota Seisan Hoshiki no Shintenkai* [The Toyota Production System's New Directions]. Tokyo: Nihon Noritsu Kyokai, 1983.

Moriguchi, Chikashi. *Nihon Keizairon*. Tokyo: Sobunkan, 1988.

Nakagawa, Yasuzo. *Toshiba no Handotai Jigyo Senryaku* [Toshiba's Semiconductor Strategy]. Tokyo: Daiyamondo, 1989.

Nakamura, Kazuo. *Shitauke Kigyo no Keiei* [Management of Subcontractors]. Tokyo: Nikkan Kogyo, 1988.

Nawa, Kotaro. *Gijutsu Hyojun tai Chiteki Shoyuken* [Technology Standards Versus Intellectual Property Rights]. Tokyo: Chuo Koronsha, 1990.

Nihon Keizai Shimbun. "Toshiba, 16.7 % Shimogata Shusei." September 1, 1992, 1.

Nihon Keizai Shimbun. "Price is the Key to Toyota's New Luxury Car for the US Market." September 14, 1994, 13.

Nihon Keizai Shimbun. "Chingin kakumei: nenkoshugi to no ketsubetsu [Wage Revolution: The collapse of seniority-based compensation]." December 13, 1995, 1.

Kiyanon no Seisan Kakushin [Canon's Production Revolution]. Tokyo: Nihon Noritsu Kyokai, 1983.

———. *Gijutsusha Kyoiku no Kenkyu* [Research on the Education of Technologists]. Tokyo: JMA Press, 1990.

———. *Kenkyusho Unei Kaseika Jitsureishu* [Studies in the Management of R & D Liveliness]. Tokyo: JMA Press, 1987.

Nikkei Business. "$1=80 Yen Factory." September 19, 1994 (No. 757), 10–22.

Nomura Research Institute. *TOSHIBA NRI Report*. Feb. 5, 1990.

Ohno, Taiichi. *Toyota Seisan Hoshiki* [The Toyota Production System]. Tokyo: Daiyamondo, 1978.

Ono, Toyoaki. *Nihon Kigyo no Soshiki Senryaku* [The Organizational Strategy of Japanese Firms]. Tokyo: Manejementosha, 1979.

Onishi, Katsuaki, and Hideitsu Ohashi. *Hitachi-Toshiba*. Tokyo: Otsuki Shoten, 1990.

Rodosogishi Kenkyukai, ed. *Nihon no Rodo Sogi 1945–80* [Japan's Industrial Strife 1945–80]. Tokyo: University of Tokyo, 1991.

Sakamoto, Kazuichi, ed. *Gijutsu Kakushin to Kigyo Kozo* [Technology Innovation and Enterprise Structure]. Kyoto: Minerva, 1985.

Sakamoto, Kazuichi, and Masao Shimotani, eds. *Gendai Nihon no Kigyo Gurupu* [Enterprise Groups in Contemporary Japan]. Tokyo: Toyo Keizai, 1987.

Shibaura Seisakusho. *Shibaura Seisakusho Rokujugonenshi* [A 65– Year History of Shibaura Engineering]. Tokyo: Bunshodo, 1940.

Shimizu, Tsutomu. *Soshiki Kaseika to Shokuba Fudo no Ikusei* [Organizational Liveliness and the Cultivation of Workplace Culture]. Tokyo: JMA Press, 1986.

Shimotani, Masahiro. "Gendai Kigyo-gurupu no kozo to kino." In Sakamoto, K., ed., *Gijutsu Kakushin to Kigyo Kozo* [Technical Innovation and Enterprise Organization]. Kyoto: Minerva Press, 1986.

Tamada, Masuo, Haruomi Akiyama, and Masayuki Kuwabara. *IC Kado Shikumi to Hirogaru Sekai* [IC Card Capabilities and Expanding Opportunities]. Tokyo: Denki Shoin, 1989.

Tokyo Shibaura Denki. *Kabushiki Kaisha Hachijugonenshi* [85 Years of the Tokyo Shibaura Electric Company]. Tokyo: Daiyamondo, 1963.

———. *Toshiba Yanagicho Yonjunen no Ayumi* [40 Years of Toshiba's Yanagicho Works]. Kawasaki: 1976.

Toshiba Corporation. *Toshiba Hyakunenshi* [A Century of Toshiba]. Tokyo: Daiyamondo, 1977.

———. *Jidoka Keikaku* [Planning for Automation]. Tokyo: Industrial Electronics Center, 1985, 31 pp.

———. *Senryaku kaigi jisshi keiei oyobi yoyakuhyo 61/shimo* [Strategic Planning Committee Meetings and Schedule for the latter half of the 1986]. Single sheet, undated, presumably 1986.

———. *Personnel Management*. Tokyo: mimeo, undated but prepared in 1986 or 1987, 12 pp, with three appendices.

———. *Sozo: Toshiba Yanagicho Kojo Gojunenshi* [Create: 50 Years of Toshiba's Yanagicho Works]. Kawasaki, 1987.

———. *Toshiba Newsletter*, Feb. 1988, No. 307, 1.

———. Toshiba factory brochures, 1990–92. Places of publication not given.

Tsuychiya, Moriaki. *Kigyo to Senryaku* [Firms and Strategies]. Tokyo: Rikuruto, 1984.

Yanagicho Works. *Heisei Nisannen TP Katsudo no Hoshin*. Photocopy, undated.

———. Labor Savings Equipment Department. *TP Progress Reports*. Kawasaki City, 1986–89.

———. Machine Tool Department. *TP Progress Reports*. Kawasaki City, 1986–89.

———. Purchasing Department. *Supplier's Management Guidelines*. Mimeo, 1987, for internal circulation only.

———. Purchasing Department. *Suppliers Management Guidelines*. Mimeo, 1988 edition.

———. "62nen/kami Kojo Undo Katsudo Hokoku." Photocopy, undated.

———. *Statistical Handbook*, updated and photocopied quarterly. June 1989 edition.

———. "3C-WG" Action Plan." *Mimeograph*, September–October 1990.

Interviews, conducted mostly in Japanese

Mr. Terunori Aiga. Toshiba Corporation. Shibaura, Tokyo, June 30, 1991, and November 5, 1994

Dr. Kohei Amo. Board Member, Toshiba Corporation. Shibaura, Tokyo, August 27, 1987.

Dr. Banri Asanuma. Economist. Kyoto, July 11, 1987.

Mr. Singo Chiba. Yanagicho Works. August 16, 1988.

Mr. Philippe Delahaye. Toshiba Systems France. Arques La Bataille, November 29, 1988.

Mr. Hitoshi Dojima. Yanagicho Works. October 23, 1990.

Mr. Susumu Domon. Yanagicho Works. August 19, 1988. And at Toshiba Systems America, Irvine, California, June 14, 1991.

Mr. Takeo Ikeda. Yanagicho Works. August 19, 1988.

Mr. Toshinori Ishizawa. Yanagicho Works. August 24, 1987.

Mr. Y. Iwatsuki. Planning Department, Yanagicho Works. August 4, 1987.

Mr. Nobuo Jikimoto. Yanagicho Works. August 3, 1987.

Dr. Akio Kameoka. Toshiba Headquarters. Shibaura, Tokyo, August 7, 1987.

Mr. Yasuo Kanesaki. Yanagicho Works. Summer 1988,

Mr. Hajime Kasahara. Yanagicho Works. August 11, 1988.

Mr. Jun Kawasaki. Toshiba Systems America. Irvine, California. June 14, 1991.

Dr. Masaharu Kinoshita. Toshiba Headquarters. Shibaura, Tokyo, June 24, 1986.

Dr. Kazuo Koike. Economist. Nagoya University. Osaka, July 13, 1987.

Mr. Matsuo "Kozan" Komai. Yanagicho Works. August 8, 1988. And December 8, 1995, in Mitchell, South Dakota.

Dr. Hiroshi Komiya. General Manager. Mitsubishi Electric Corporation, LSI Research and Development Laboratory, Itami City, Japan, February 8, 1991.

Mr. Yoshio Kondo. Head, Training Department. Yanagicho Works. Spring 1991.

Mr. Junji Kubota. Yanagicho Works. June 16, 1986.

Mr. Akira Kuwahara. Toshiba Headquarters. Shibaura, Tokyo, August 8, 1991.

Mr. Hiroyuki Matsuda. Yanagicho Works. August 17, 1988.

Mr. Kenichi Mori. Toshiba Headquarters. Shibaura, Tokyo, April 21, 1989.

Mr. Shuji Mori. Ricoh R & D Center. Yokohama, Japan, April 20, 1989.

Mr. Hiroshi Nakanishi. Toshiba Systems France. Arques La Bataille, November 29, 1988.

Mr. Makoto Namikawa. Yanagicho Works. August 25, 1988.

Mr. Sirou Okajima. Yanagicho Works. August 18, 1988.

Mr. Yuji Osada. Yanagicho Works. August 6, 1987, and April 4, 1988.

Mr. Yoshitaka Ota. Yanagicho Works. August 4, 1987, January 12, 1988, and August 20, 1988.

Mr. Eiichi Sano. Toshiba Headquarters. Shibaura, Tokyo, June 24, 1986.

Mr. Yoshio Shiba. Toshiba Systems America, Irvine, California. 1/26/90.

Mr. Tomonobu Shibamiya, Head, Manufacturing Engineering, August 3, 1987.

Mr. Tsutomu Shimizu. Toshiba Headquarters. Shibaura, Tokyo, June 27, 1985.

Mr. Seiichi Shimoi. Toshiba Headquarters. Shibaura, Tokyo, October 5, 1990.

Mr. Hiro Shogase. Toshiba Headquarters. Shibaura, Tokyo, August 29, 1993.

Mr. Kazuiti Sumiyoshi. Yanagicho Works. June 10, 1986 and February 25, 1991.

Mr. Yasuo Suzuki. Ome Works. Ome City, June 28, 1986.

Dr. Seiichi Takayanagi. Senior Advisor. Toshiba Headquarters. Shibaura, Tokyo, August 19, 1994. And at public presentation, Almasa, Sweden, September 20, 1994.

Mr. Masuo Tamada. Yanagicho Works. April 4, 1987. And at IC Card Systems, Yanagicho Works, February 20, 1990.

Mr. Masaharu Tanino. Toshiba Headquarters. Shibaura, Tokyo, July 7, 1987, August 11, 1987, August 18, 1987, August 27, 1987.

Mr. Uchida, Yanagicho Works. Kawasaki City, June 19, 1986.

Mr. Kazuichi Wada. Yanagicho Works. August 18, 1988.

Mr. Sadakazu Watanabe. Toshiba Headquarters. Tokyo, April 26, 1989.

Mr. Roger N. Wiewel. Senior V. P. for Product Development. MacMillan Bloedel. January 30, 1991 in Tokyo.

Mr. Tetsuya Yamamoto. Toshiba Headquarters. Shibaura, Tokyo, October 16, 1990, August 7, 1991, April 3, 1992, and August 26, 1993.

Mr. Masanori (George) Yamoto. Toshiba Systems America, Irvine, California. May 30, 1995.

Mr. Yoshie Yoshida. Horikawa Works. Kawasaki City, June 26, 1985.

Mr. Hitoshi Yoshii. Purchasing Department. Yanagicho Works. August 9, 1988.

Mr. Yoshi Yoshinaga. Yanagicho Works. August 18, 1988.

Yanagicho Works Line and Staff Organization

Early 1990s

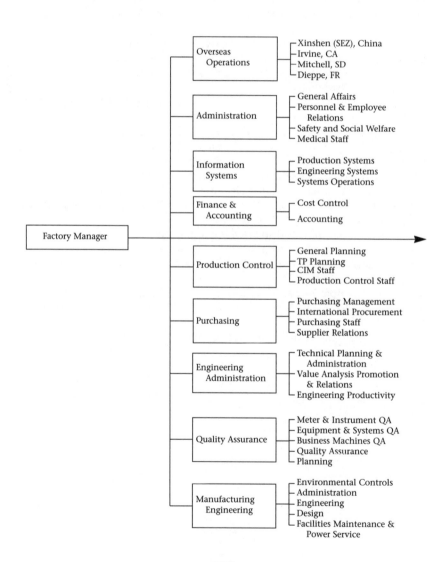

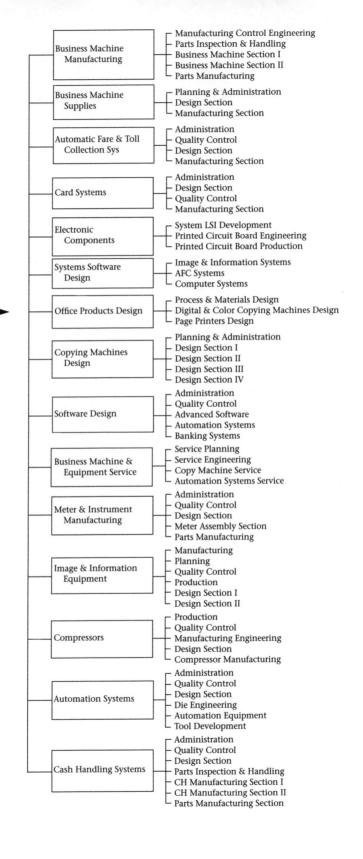

Business Machine Manufacturing
- Manufacturing Control Engineering
- Parts Inspection & Handling
- Business Machine Section I
- Business Machine Section II
- Parts Manufacturing

Business Machine Supplies
- Planning & Administration
- Design Section
- Manufacturing Section

Automatic Fare & Toll Collection Sys
- Administration
- Quality Control
- Design Section
- Manufacturing Section

Card Systems
- Administration
- Design Section
- Quality Control
- Manufacturing Section

Electronic Components
- System LSI Development
- Printed Circuit Board Engineering
- Printed Circuit Board Production

Systems Software Design
- Image & Information Systems
- AFC Systems
- Computer Systems

Office Products Design
- Process & Materials Design
- Digital & Color Copying Machines Design
- Page Printers Design

Copying Machines Design
- Planning & Administration
- Design Section I
- Design Section II
- Design Section III
- Design Section IV

Software Design
- Administration
- Quality Control
- Advanced Software
- Automation Systems
- Banking Systems

Business Machine & Equipment Service
- Service Planning
- Service Engineering
- Copy Machine Service
- Automation Systems Service

Meter & Instrument Manufacturing
- Administration
- Quality Control
- Design Section
- Meter Assembly Section
- Parts Manufacturing

Image & Information Equipment
- Manufacturing
- Planning
- Quality Control
- Production
- Design Section I
- Design Section II

Compressors
- Production
- Quality Control
- Manufacturing Engineering
- Design Section
- Compressor Manufacturing

Automation Systems
- Administration
- Quality Control
- Design Section
- Die Engineering
- Automation Equipment
- Tool Development

Cash Handling Systems
- Administration
- Quality Control
- Design Section
- Parts Inspection & Handling
- CH Manufacturing Section I
- CH Manufacturing Section II
- Parts Manufacturing Section

Index